Internet of Things

Technology, Communications and Computing

More information about this series at http://www.springer.com/series/11636

Zoltan Papp · George Exarchakos
Editors

Runtime Reconfiguration in Networked Embedded Systems

Design and Testing Practices

Springer

Editors
Zoltan Papp
TNO
The Hague
The Netherlands

George Exarchakos
Technical University Eindhoven
Eindhoven
The Netherlands

ISSN 2199-1073 ISSN 2199-1081 (electronic)
Internet of Things
Technology, Communications and Computing
ISBN 978-981-10-0714-9 ISBN 978-981-10-0715-6 (eBook)
DOI 10.1007/978-981-10-0715-6

Library of Congress Control Number: 2016933197

Printed on acid-free paper

This Springer imprint is published by Springer Nature
The registered company is Springer Science+Business Media Singapore Pte Ltd.

To Zoltan Papp, in memoriam.

Zoltan was a wonderful, inspiring, and good-hearted friend, who gave shape and content to this book through guidance, generous advices, and insightful discussions. As he would himself explain: "It is not fate, it is entropy".

Foreword

More than 7 years ago I started working in the field of the Internet of Things (IoT), even if I did not know exactly what it was meant for, and probably nobody knew at that time, as this novel concept had still to find the exact definition and role. Nonetheless, I was fascinated by the fact that the real world could have virtual representation allowing for the realization of an unimaginable range of possible applications for the benefit of the people quality of life. Like many other colleagues, we imagined *IoT as a solution to integrate* a large number of heterogeneous and pervasive objects with different connecting and computing capabilities, with the aim of providing a view of the status of the physical world through the network. These technologies were expected to make possible the realization of new services to end users in disparate fields, from the environment monitoring to the industrial plants running, from city management to house monitoring. Since that time, the variety of devices capable to send data about the physical environment to the Internet and to take local and autonomous decisions has been ever-increasing: these now range from dummy entities capable of providing just their positions (through attached tags) to objects capable of sensing the status of an environment, processing the data, and sending to external systems, if believed meaningful. Clearly, *networked embedded systems* are key components in this picture as they allow to introduce rapid and local decisions in the environment and to implement adaptive solutions.

Leveraging on my background in quality of service (QoS) management in next generation networks, at the beginning I started identifying what I believed were the most challenging networking issues for some selected IoT applications, i.e., optimizing the configuration of multi-homed wireless sensor networks. I then started modeling some components of the target systems with particular attention to the impact of the proposed solutions on the final performance in terms of QoS. I soon discovered that the approach I followed, like many other colleagues with a strong focus on the networking aspects, had two common limitations. First, it produced solutions that were incomplete, as they did not consider issues not directly associated to the connectivity, such as those related to the functioning of real-time processing systems, which might have introduced not negligible delays and additional partial or complete system failures. Second, it had the limitation related to the adaptability of the complete systems which was neglected; this instead is a must for components of IoT systems that need to evolve over the time along with external systems with which it interacts.

Much of my progress in this field happened, also thanks to the DEMANES (Design, Monitoring and Operation of Adaptive Networked Embedded Systems) project, which we have devised in 2012 with an enthusiastic team of colleagues from European research centers and industries, and which is still running. It is during this experience that I learnt a world of key methodologies and tools required for an effective design and management of networked embedded systems. Indeed, designing effective and deploying powerful IoT systems with strong links with the physical world requires a deep knowledge of modeling language for complex systems (often systems of systems) with the capabilities to describe the disparate properties and behaviors of the components under analysis. Often hostile environments are encountered and in this case it is not always simple to reuse models designed by others and often everything has to be done from scratch. Runtime reconfiguration capabilities (covering self-organization and self-optimization) have also often to be realized to extend the operational envelope of the system. After the first design has been accomplished, it is then necessary to determine if the system performs properly in different scenarios or goes astray into some state that does not correspond to a viability zone. This is the objective of the test and validation activity, which is aimed at understanding the state of the system after an adaptation action and also to understand the quality of the adaptation action. All these validation activities have to go through a rigorous and a well-defined set of rules. It also needs an appropriate framework, which is usually comprised of three main elements: a test coordinator, a system monitor, and an analyzer. The latter component contains the testing algorithms that gather the system variables from the monitors to verify the system status at runtime in terms of viability zones and adaptivity, according to the instructions set by the test coordinator.

This book covers all these aspects in a compressive way representing a powerful tool for the colleagues that have to deal with the design and deployment of networked embedded systems, whether they are a part of complex IoT solutions or not. It focuses on the special issues in designing and deploying runtime reconfiguration functionalities in real-time context addressing the model-based engineering, coping with complexity in runtime, managing the design-time—runtime trade-off, etc. The book follows a paragmatic approach for solving this complex design problem: it helps in understanding the applicability of the presented theories and tools by discussing representative use cases.

Cagliari Luigi Atzori
June 2015

Preface

In order to address todays societal challenges (e.g., environmental management, intelligent mobility, safety/security, sustainable urban living, etc.) monitoring and control of large-scale, complex, dynamical systems are essential technologies. These monitoring and control systems realize sophisticated signal processing algorithms (state estimation, pattern recognition, optimal control, decision-making, etc.) in distributed configurations and many times are deployed in difficult-to-access, hostile environments. Under these circumstances failures, changes in operational conditions and changes in user needs are not exceptions but should be considered as nominal operation. Runtime reconfiguration capabilities (covering self-organization, self-optimization) have to be realized to extend the operational envelope of the system. Runtime reconfiguration enables the system to cope with changing system configurations without re-programming the existing setup (i.e., supporting easy deployment), to support mobile subsystems (dynamic sensor management), to cope with environmental changes (e.g., changing network capacities) and to adapt to changing system goals (user needs) during operation. The realization and operation of runtime reconfigurable systems poses new challenges in every stage of the system's life cycle.

In the book the design and testing of large-scale, distributed signal processing systems are considered with special emphasis on systems architecture, tooling, and best practices. Distinguishing features of the targeted systems are serious resource constraints (e.g., processing capabilities, communication, energy supply), and the presence of demanding nonfunctional requirements such as timeliness, robustness, lifetime, and the capability of handling system "evolution". All along a strict model-based process is followed and thus the architecture modeling, model checking, model-based evaluation, and model-based design optimization play a central role in the book. After summarizing the conceptual/theoretical foundation of the methodology, a systems engineering process and a matching tool suite is described. In order to assure the landing of the abstract concepts, we shall use a number of nontrivial examples, taken from real-world cases, to illustrate the application of theory to practice.

The book is intended to be used as a cookbook for designers and practitioners of complex embedded systems: though it covers the relevant fundamentals for designing and testing of large-scale networked embedded systems, it also proposes systems engineering processes, architectural templates (design patterns), and practical testing/validation approaches and thus it serves as a guideline for the system designer.

The book captures the expertise of several authors who have worked for years at the interface between theory and engineering in a variety of topical issues of embedded systems. Since 2012 these authors have come to work together under the umbrella of the European project ARTEMIS DEMANES (Design, Monitoring and Operation of Adaptive Networked Embedded Systems) that has at its heart the ambition to define a methodology and to realize tool chain for high-level support for designing, implementing, testing, and deploying real-world runtime reconfigurable applications. This book collects the most relevant developments by DEMANES members in the area of self-configurable embedded networks and systems. This manuscript is presenting seven complementary topics relevant to the requirements specification, the off-line design methodology, the services and tooling used, the validation and verification framework, the validation tools and, finally, the applicability to real-world pilots of large-scale runtime-adaptive networked embedded systems.

In Chap. 1, a multi-aspect modeling language is introduced that allows system designers to model the architecture of large-scale networked systems from different aspects and contains concepts to model runtime reconfiguration at design-time. The proposed architecture for modeling runtime reconfiguration consists of primary tasks in one layer and secondary management tasks in another layer. Special reconfiguration primitives allow the description of four types of reconfiguration: re-parameterization, re-instantiation, rewiring, and relocation. The modeling language is accompanied by a modeling methodology based on MAPE-K and uses feedback loops in the system model to model runtime reconfiguration. System-level characteristics, i.e., KPIs are introduced allowing system designers to evaluate system designs and pick the most promising one. Special attention is paid to the fact that the availability of a runtime redesign capability in a system requires KPIs to be derived and evaluated at runtime. Only then can the redesign be guided properly.

Chapter 2 discusses design methodologies that are specific for adaptive networked embedded systems. This chapter discusses design-time/runtime trade-offs, design patterns for reconfigurable real-time monitoring and control, runtime design space exploration of adaptive systems and a systems engineering process for runtime reconfigurable systems. We draw guidelines for all stages of the architectural process and help system and software designers in choosing wisely specific algorithms and techniques. In summary, we showcase state-of-the-art design techniques that are directly addressing the adaptivity.

Chapter 3 discusses how to build up a model-oriented tool chain and its most important activity flows: model editing, model visualization and transformation; model validation and evaluation; and finally, (semi-)automatic system implementation based on the system model. In addition, for each activity emphasis goes also to the specific tool support which is necessary for the design of adaptive/reconfigurable systems.

Chapter 4 recognizes and analyzes the need for runtime validation and verification of large-scale adaptive networked embedded systems (ANES). It is not feasible, during the development stages, to anticipate all the possible operating conditions that the system may face in a real environment. This is because some information about the execution context and the system itself can be available only once the system has been deployed. Thus, in order to correctly assess the effectiveness, efficiency and robustness of ANES, it is required to verify that the system correctly adopts the proper adaptation mechanisms in response to the context changes as well as to check the quality of such adaptations. The focus of the chapter is to discuss about the needs for employing runtime verification and validation of ANES and the main challenges and requirements for its implementation. Moreover, it presents a reference framework that supports developers in testing adaptive systems at runtime. One of its key features is the capability to emulate certain realistic conditions through synthetic data, which is useful to check the system's behavior under specific and controlled situations.

Chapter 5, symmetrically to Chap. 3, is a thorough analysis of the state-of-the-art tooling available for validating and verifying at runtime large-scale networked embedded systems. In cases where the long-term functioning of the systems is of interest, the systems quality should prove very high and, therefore, proper validation and verification practices are required. Given the lack of relevant tools that deal with testing runtime self-adaptive systems, this chapter proposes the implementation of a V&V framework introduced by Chap. 4, by merging several already known tools. First, it gives an understanding of ways to quantify and predict the reliability of large-scale distributed systems. Second, key performance indicators of the self-adaptive systems are identified from monitoring techniques and third, the test cases are formalized in a structured form. Finally, the chapter presents two test cases as examples of a system working under normal operation conditions as well as under induced conditions, based on real-life implementations. Execution of the test is led by a test coordinator for which JSON notation is used, and then the interpretation and testing is carried out in a C++ toolbox where the monitoring and testing algorithms reside.

This last chapter describes a real self-adaptive system carried out using the DEMANES tool chain based on the foundations of the DEMANES tool chain described from previous chapters. This chapter focuses on the design and implementation stages of a real use case development, a pilot. The use case under study is a subsystem called cargo monitoring system (CMS), which monitors the state of the container cargo and push all the data to a back office infrastructure for further processing. The containers can be on a truck, a train, or any other appropriate transportation means, or stacked in a container terminal or a cargo ship. A WSN will measure physical magnitudes (temperature, humidity and so on) inside a container and will forward data to others processing nodes in the CMS network. The CMS has to meet several self-adaptive requirements. For instance, the parameters of the CMS elements that monitor the container cargo state are reconfigured accordingly to adapt to internal or external changes (e.g., a low battery level or a container temperature out of the adequate bounds), and the CMS adapts the WSN nodes power transmission to save energy while providing an acceptable quality of service.

Delft, The Netherlands Zoltan Papp
Eindhoven, The Netherlands George Exarchakos
June 2015

Acknowledgments

Nearing its closure, DEMANES is an ARTEMIS-IA project and the catalyst that made this book possible. We, as editors of this volume, would like to express our appreciation to the authors for their contributions and to all DEMANES partners who set the ground for building this book. Throughout the project and book editing the project officer of DEMANES, Patrick Vandenberghe, has consistently and constructively provided valuable feedback and guidance. Thanks to his support and motivation, this book came to a completion. Special thanks also go to ARTEMIS-IA for funding DEMANES and providing the means for collaboration between partners with complementing interests. Finally, we would like to name the DEMANES partners out of appreciation for the experiences fed into this book and made the content relevant and applicable.

Czech Republic
Czech Technical University in Prague
Merica Sro

Finland
University of Eastern Finland
Convergens Oy
Mega Elektroniikka Oy

Italy
Akhela S.r.l.
Vitrociset S.p.A.
Universit degli Studi di Cagliari
Selex ES S.p.A.
Politecnico di Torino
Azienda per la Mobilit del Comune di Roma S.p.A.

The Netherlands
Technische Universiteit Delft
Cooperatie DevLab
NXP Semiconductors B.V.
Philips Electronics & Lighting B.V.
Technische Universiteit Eindhoven
Nederlandse Organisatie voor Toegepast Natuurwetenschappelijk Onderzoek

Spain
Hi Iberia Ingeniería y Proyectos S.L.
Instalaciones Inabensa S.A.
Integrasys S.A.
Information & Image Management System S.A.
Acondicionamiento Tarrasense Asociación
Fundación Tecnalia Research & Innovation
Universidad Politécnica de Madrid

Contents

Contributors

Claudio Bacchiani TU/e, Eindhoven, The Netherlands

Julio Oliveira de Filho TNO, The Hague, The Netherlands

Jan de Gier TNO, The Hague, The Netherlands

Julio de Oliveira Filho TNO, The Hague, The Netherlands

Raúl del Toro Matamoros UPM, Madrid, Spain

Vicente Hernández Díaz UPM, Madrid, Spain

Stefano Galzarano TU/e, Eindhoven, The Netherlands

Paola Jaramillo TU/e, Eindhoven, The Netherlands

Roshan Kotian TU/e, Eindhoven, The Netherlands

Jean-Paul Linnartz TU/e, Eindhoven, The Netherlands

Zoltan Papp TNO, The Hague, The Netherlands

Andrei Pruteanu TU Delft, Delft, The Netherlands

Yolanda Rieter-Barrell TNO, The Hague, The Netherlands

Přemysl Šucha Czech Technical University, Prague, Czech Republic

Aly A. Syed NXP, Eindhoven, The Netherlands

Roman Václavík Czech Technical University, Prague, Czech Republic

Willem van Driel Philips Lighting, Eindhoven, The Netherlands

Wijnand van Kooten Philips Lighting, Eindhoven, The Netherlands

Coen van Leeuwen TNO, The Hague, The Netherlands

Teus Vogel TNO, The Hague, The Netherlands

Acronyms

3G	Third Generation of Mobile Communications Technology
AADL	Architecture Analysis and Design Language
ALT	Accelerated Life Testing
ANES	Adaptive, Networked and Embedded Systems
ARM	Advanced RISC Machines
CLUE	Cargo Life Update Event
CMS	Cargo Monitoring System
CoAP	Constrained Application Protocol
CPU	Central Processing Unit
CR	Coverage Region
DDR	Double Data Rate type 3
DEMANES	DEsign, Monitoring and operation of Adaptive Network Embedded Systems
DSL	Domain-Specific Language
eMMC	embedded MultiMedia Card
FIFO	First-In First-Out
FLD	Fuzzy Logic Device
FMEA	Failure Mode and Effect Analysis
FR	Functional Requirements
GA	Genetic Algorithm
GME	Generic Modeling Environment
GT	Ground Truth
HW	Hardware
IEEE	Institute of Electrical and Electronics Engineers
ISO	International Standardization Organization
ITU	International Telecommunication Union
JME	Java Micro Edition
JSON	Java Script Object Notation
JVM	Java Virtual Machine
KPI	Key Performance Indicator

MAPE-K	Monitor Analyze Plan Execute Knowledge
microSD	micro Secure Digital (card)
MW	MiddleWare
ND	Node Degree
NES	Networked Embedded System
NFR	NonFunctional Requirements
nrFLOps	number of FLoating point operations
nrIOps	number of Input Operations
OSGi	Open Services Gateway initiative
PF	Primary Functionality
PHP	PHP: Hypertext Preprocessor
RAM	Random-Access Memory
REST	Representational State Transfer
RF	Reconfiguration Functionality
RM-ODP	Reference Model—Open Distributed Processing
ROM	Read-Only Memory
RSSI	Received Signal Strength Indicator
SAE	Society of Automotive Engineers
SE DSIG	System Engineering Domain Special Interest Group
SLA	Service Level Agreement
SLO	Service Level Objective
SOA	Service-Oriented Architecture
SQL	Structured Query Language
SW	Software
SysML	Systems Modeling Language
UML	Unified Modeling Language
USB	Universal Serial Bus
V&V	Verification and Validation
VZ	Viability Zone
WiFi	Wireless Fidelity (wireless connectivity standard)
WSN	Wireless Sensor Network
WSAN	Wireless Sensors and Actuators Network
XML	W3C eXtended Markup Language
XSD	W3C XML Schema Definition Language

Chapter 1
Model-Based Engineering of Runtime Reconfigurable Networked Embedded Systems

Coen van Leeuwen, Yolanda Rieter-Barrell, Zoltan Papp, Andrei Pruteanu and Teus Vogel

Abstract Today's societal challenges, such as sustainable urban living and public safety and security require monitoring and control solutions for large-scale complex and dynamical systems. The distinguishing features of these systems are serious resource constraints, demanding non-functional requirements such as robustness, timeliness, lifetime and the capability of handling system evolution through runtime reconfiguration. In this chapter, a multi-aspect modeling language is introduced that allows system designers to model the architecture of large scale networked systems from different aspects. This modeling language introduces innovative concepts to model runtime reconfiguration at design-time. The proposed architecture for modeling runtime reconfiguration consists of primary tasks in one layer and secondary management tasks in another layer. Special reconfiguration primitives allow the description of four types of reconfiguration: re-parameterisation, re-instantiation, rewiring and relocation. The modeling language is accompanied by a modeling and design methodology (inspired by the MAPE-K technique [1]) and uses feedback loops in the system model to realize runtime reconfiguration. This chapter also proposes Key Performance Indicators (KPIs) that allow designers to quantify the "quality" of the system designs and pick the most promising one. Special attention is paid

C. van Leeuwen · Y. Rieter-Barrell (✉) · Z. Papp · T. Vogel
TNO, Oude Waalsdorperweg 63, The Hague, The Netherlands
e-mail: yolanda.rieter@tno.nl

C. van Leeuwen
e-mail: coen.vanleeuwen@tno.nl

T. Vogel
e-mail: teus.vogel@tno.nl

A. Pruteanu
TU Delft, Postbus 5, Delft, The Netherlands
e-mail: a.s.pruteanu@tudelft.nl

Z. Papp and G. Exarchakos (eds.), *Runtime Reconfiguration in Networked Embedded Systems*, Internet of Things, DOI 10.1007/978-981-10-0715-6_1

to the fact that the availability of a runtime reconfiguration (i.e. re-design capability) in a system requires KPIs to be derived and evaluated at runtime as a precondition for guiding the reconfiguration process.

1.1 Introduction

Addressing today's societal challenges (e.g. sustainable urban living, renewable energy, public safety/security) requires monitoring and control solutions for large-scale, complex, dynamical systems. These monitoring and control systems realize sophisticated signal processing algorithms in distributed configurations, and many times they are deployed in difficult to access, hostile environments. Runtime reconfiguration capabilities—providing self-organization and self-optimization—have to be realized to extend the operational envelope of the system. The design of runtime reconfigurable systems poses new challenges both for the signal processing and system architecture aspects.

Model-based design is an accepted methodology for designing systems and quantitatively assessing them. Standardized modeling languages are available for modeling system architectures. Commercial tools are available for modeling, analysis and optimization of the system architecture designs. However, these tools and their underlying formalisms either focus on a single application domain or have limits on capabilities and scalability. Moreover, most tools enable system designers for making static designs but do not provide concepts for modeling runtime reconfiguration.

A multi-aspect modeling language is introduced in this chapter. This language allows system designers to model the system architecture from different aspects (e.g. functional, physical, etc.) and contains concepts to model runtime reconfiguration at design-time. Distinguishing features of the targeted systems are serious resource constraints (e.g. processing capabilities, communication, energy supply), and the presence of demanding non-functional requirements, such as timeliness, robustness, lifetime and the capability of handling system evolution. The reconfiguration can be considered as a technique to implement this evolution. The reconfiguration can be carried out in different stages of the systems life cycle, such as design time, load time, initialization time and runtime. Here special emphasis is given to the runtime reconfiguration variant dedicated to signal processing type of applications.

First, four modeling aspects are introduced: the task, behavioral, physical and mapping aspects. The most important characteristics of each aspect is introduced and the interrelations between the various aspects is addressed briefly.

In order to simplify the modeling and management of large scale networked embedded systems in a concise and compact manner without loosing flexibility, the concept of dynamic instantiation is introduced in the language.

The modeling language includes explicit concepts associated with particular reconfiguration architectures. This architecture comprises a reconfiguration management layer, which monitors and controls the primary functionality of the system. Four types of reconfiguration primitives have been identified: re-parameterization, re-instantiation, rewiring and relocation. Modeling concepts are introduced to support all four reconfiguration types.

The chapter is concluded with a modeling methodology tailored to runtime reconfigurable systems accompanied by examples.

1.2 Multi-aspect Modeling for Networked Embedded Systems

The large scale nature of networked embedded systems (NESs) influences the complexity of the architecture design from two perspectives: the architecture description and the architecture assessment. Challenges for the architecture assessment arise, for example, when dimensioning and characterizing communication channels, or when profiling power consumption and dependability of the system, among others. To cope with architecture assessment challenges, the designer has to employ adequate analysis tools relying on formal specifications of relevant aspects of the design (i.e. models).

Complex systems consist of entities that interact with each other to produce the behaviour of the system as a whole. An important characteristic of a complex system is that the properties and behaviour of the whole are emergent; that is the system level behaviour cannot simply be inferred from the properties and behaviour of the components. Many relatively simple entities interact in relatively simple ways to give rise to emergent phenomena that could not be "visible" from the definition of the entities.

The performance of interactive systems is determined in relation to the context in which the system performs its intended roles. A system that performs well in one context may not perform well in other contexts. While a system's context may dynamically change, traditionally systems have static designs that allow operation within a limited range of context variations. The book targets systems that are sensitive to changes in the environment they operate in, and are able to adapt to a large range of contexts. The system modeling methodology used to design the system architecture must be able to address adaptivity of the system to a changing environment.

Existing modeling and assessment tools do not meet the system designer's needs to face the challenges of designing large scale runtime reconfigurable NESs. In the following sections modeling language concepts will be introduced which better meets those needs.

1.2.1 Related Work

Many approaches to multi-aspect modeling can be found it the literature. Some of them target specific application domains, while others are more general purpose.

For example, RM-ODP (Reference Model-Open Distributed Processing), a reference model introduced in the eighties as the result of a cooperative effort by the ISO (International Standards Organization) and ITU-T (International Telecommunication Union) [2–5]. RM-ODP provides a framework through which analyzing, describing and specifying a system from different perspectives, called viewpoints. Each of these viewpoints tends to satisfy a different audience concerned with specific aspects of the system. Associated with each of the viewpoints, a specialized language is defined that includes the vocabulary and the expressions of the particular audience to which it is addressed.

Another example is the Architecture Analysis and Design Language (AADL), which was standardized by the Society of Automotive Engineers (SAE) [6]. AADL defines a language for describing both the software architecture and the execution platform architectures of performance-critical, embedded, real-time systems. An AADL model describes a system as a hierarchy of components with their interfaces and their interconnections. Properties are associated to these constructions. AADL components fall into two major categories: those that represent the physical hardware and those representing the application software.

SysML is a general-purpose modeling language for systems engineering that supports the specification, analysis, design, verification and validation of a broad range of complex systems, including hardware, software, information, processes, personnel and facilities [7]. It uses a subset of UML 2.1 and provides additional extensions needed to fulfill the requirements for the modeling language specified by the SE DSIG (Systems Engineering Domain Special Interest Group) of the OMG.

However, all these multi-aspect modeling approaches lack the possibility to describe the dynamic reconfiguration aspects of a system. The DEMANES multi-aspect modeling language differs from the previous approaches, in that it contains language constructs that can effectively capture the dynamic reconfiguration behaviour of a system. This will be further discussed in Sect. 1.4

1.2.2 System Models

Using models in modern engineering disciplines is a common practice. This mainly has to do with the fact that the use of models is very attractive from an efficiency and economic point of view. Models are faster and cheaper to construct and easier to manipulate than the real (physical, full-scale) artifacts they describe. In experimental setups, models can be subjected to stimuli and conditions that would not be feasible or just be too dangerous to carry out with the real artifact.

Nowadays, computer models are used in many engineering disciplines to analyze and predict the behaviour of the systems they describe. Engineers use dedicated programs to interactively create and manipulate the models and to simulate/evaluate the behaviour of the systems using various test scenarios.

We can make a distinction here between informal models and formal models. Informal models are built from elements and use constructs that are at best informally documented. Alternatively, formal models are constructed using modeling elements, follow construction rules, and conform to constraints that are strictly and explicitly defined in a formal language and associated with well-defined execution semantics. The modeling elements constitute the 'alphabet' of the formal modeling language, while the possible relations and construction rules between the modeling elements are the language syntax.

Informal models are sometimes used to sketch preliminary design ideas or to convey design ideas among designers. The semantics of the modeling elements is not strictly defined and mostly left to the individual interpretation of designers. Often, this leaves room for both ambiguities and inconsistencies between interpretations among different designers. At best, these models are stored in a document oriented way.

Formal models, on the other hand, can be checked for language conformance, internal consistency and completeness, and ultimately can be used to analyze the properties of the systems they describe. When these models are made machine-readable, software tools can be constructed to manipulate the models and to perform automated operations on them like checking the consistency, constraints and completeness of the model and execute internal model transformations. This offers a great benefit, since carrying out these operations by humans would be too error-prone or could simply be not feasible due to the size or complexity of the model.

Once the semantics of the modeling elements, properties and relations are properly defined, a system designer can use the models to analyze certain properties of the system they describe and predict system behaviour. In some disciplines, and when the models present sufficient semantic content, we can use the models to perform system synthesis, that is, automatically construct (generate) parts of the system's realization. Note, that the semantics of a modeling language is ideally defined explicitly as well. Often, the language semantics is only informally described and fixed by the implementation of the analysis tools, simulators and synthesis tools acting on the models.[1]

Formal models can be given a textual as well as a graphical representation. Mostly, designers prefer graphical presentations of a model, typically as a set of diagrams. Graphical metaphors are chosen that closely match the abstractions used by the designer when designing the system. Also, various relations between system components can be shown explicitly in the diagrams. A mixed form of model presentation is also possible, in which parts of the model are expressed using graphical elements and other parts are specified using a textual formalism.

It is worth noting that beside the advantages of using formal models to describe systems, there can be pitfalls as well. Formal models capture the properties of a system by using a limited set of abstract modeling primitives. The level of detail as well as the level of accuracy with which system properties are described has its

[1]For large and complex systems, the model can now also be used as a specification for the independent realization of system components in parallel by different teams and possibly at different locations. After the realization of the components is complete, the model can be used to verify their construction and properties and subsequently, the components can be integrated into the final system.

limits. Also, certain aspects of the real system can be completely left out, e.g. certain physical interactions between parts of the system that will exist in reality, may not be reflected at all by the model. Mostly, these aspects are intentionally omitted to keep the models from becoming overly complex and because the associated component interactions do not play a dominant role in system behaviour.

As a consequence, model based system analysis will only provide approximations for system characterization and system behaviour. In order to improve the accuracy of the analysis, certain property values of the modeling elements may need to be verified/calibrated using the real physical components as a reference.

1.2.3 Multi-aspect Modeling

When using model-based design methodologies formal modeling languages can be defined such that they allow the description of a target system using multiple system aspects.

A *system aspect*, or *system view*, is a way to look at or describe a system as a whole. Each system aspect has its own associated semantic domain and can provide an exhaustive description of the system, but only from that particular point of view. Different groups of users of a system may consider completely different aspects of that system. For example, an accounts clerk will have a completely different view of the companies administrative system than its system developer.

Using multi-aspect modeling framework provides a number of advantages:

- It allows to capture the different ways separate groups of users of a system view that system. Each group of users or stakeholders, has its own concerns with respect to the system to be realized, possibly expressed as a set of requirements in a particular semantic domain (e.g. responsiveness, throughput, energy consumption, dependability, etc.).
- The development of a system typically involves the cooperation of multiple design disciplines. Each discipline will typically be addressed by only a subset of experts of the team.
- The different system aspects are not completely independent, since each aspect ultimately describes the same physical artefact. Each system aspect only offers one angle on the system's realization. Together, they form an integrated set and collectively offer a complete specification for the realization of the system.
- Design decisions made in one discipline can have consequences in other disciplines. Using multi aspect models and the relevant analysis tools make this kind of interdisciplinary design trade-off manageable. (However, it remains a challenge to engineer a multi aspect modeling language such, that the semantics of the different aspects are as orthogonal as possible, that a single point of definition is achieved.) A multi-aspect modeling language will improve the usability of the models and offer greater flexibility in the exploration of design alternatives as the different system aspects in the model can be manipulated more independently.

- It is a very powerful means to reduce model complexity: it enables designers to focus on only one system aspect at a time. By starting at the more abstract system aspects, the system design effort can progress in stages, where at each subsequent stage more detail of the system components is added and more component interrelations are captured by the model.
- The models subsequently can be used for many purposes that can aid a system architect, such as automatic code generation, design optimizing, system evolution etc. [8]. The interaction between the different models then plays a crucial role in the design process.

The following section introduces the DEMANES modeling language from a semantic point of view. The DEMANES modeling language is a multi-aspect formal modeling language with both graphical and textual presentation. The language consists of several concepts to support the design of large scale NESs and runtime reconfiguration. A comprehensive description of the modeling language can be found in [9], where for each of the modeling aspects, the abstract syntax as well as the graphical syntax is presented.

The DEMANES modeling language supports four modeling aspects: the task aspect, the behavioral aspect, the physical aspect and the mapping aspect. Each of these aspects will now be discussed in more detail.

1.2.4 The Task Aspect

The task aspect is covered by a task model. It describes an application as a collection of communicating tasks. A task represents a sequential application function that will be executed in parallel with other tasks (i.e. the task model describes also the concurrency features of the system). Therefore, the combined collection of all tasks in the task model of an application reflects the maximum possible concurrency that can exist in the modeled system (constrained by task synchronization and resource access/usage). What a task should carry out as a function and how it is carried out is defined in the behavioral model associated with the task.

Tasks coordinate the work by communicating and synchronizing with each other, i.e. tasks are interconnected. The connections are called links and have flow semantics: flow of data messages.

Tasks have an interface consisting of a set of ports, i.e. these are the terminals of the links connecting the tasks. The ports are the only part of a task that is made visible to the outside world: the rest of a task is hidden from the outside world. We distinguish between data input ports and data output ports. All ports have a unique name within the context of the associated task. The input ports have a message queue (FIFO type); this can be used to model different types of synchronization behaviour. Supported are the asynchronous communication (queue size > 0) and synchronous communication (queue size $= 0$).

Inside the task the ports are visible, and can be associated with communication operations that are defined in the behavioral model. This way, ports can be used to separate the behavioral specification from the inter-task communication topology and synchronization specification.

To facilitate modeling of large scale systems, tasks can be grouped into *task groups*. Task groups are just a syntactic notation (collections of tasks) and does not influence the execution of tasks (i.e. task groups are "flattened out" from execution point of view), and have an interface consisting of the outer ports. This modeling mechanism allows a designer to organize the model content in a hierarchical way. Hierarchical modeling will reduce the complexity of a model and can capture valuable structural design information. The tasks at the bottom of the hierarchy represent the sequential parts of the system.

Example
Figure 1.1 shows the task model for a very simple application. The task model contains a task group and a task named respectively 'remoteDataAcq' and 'collectSensorData'. The tasks are visualized as rectangles, with the name of the task shown in a header. Each task has one port. Input ports are represented by a small box with a dot inside. Output ports are represented by a small box with a cross inside. Task 'remoteDataAcq' has an output port p_{out} while task 'collectSensorData' has an input port p_{in}. The two tasks communicate/synchronize by using a dataflow named 'sensorData'. The dataflow arrow connects the output port to the input port.

Finally, a replicator object is added to the task model on the task group 'remoteDataAcq'. The replicator is visualized by a triangular shape and has a name (i) and a cardinality (4). The semantics associated with a replicator indicates the acted-upon objects are to be instantiated multiple times. The exact number of instances is specified by the cardinality property of the replicator, which can be an expression. Replicators can occur everywhere in the task aspect, the physical aspect and

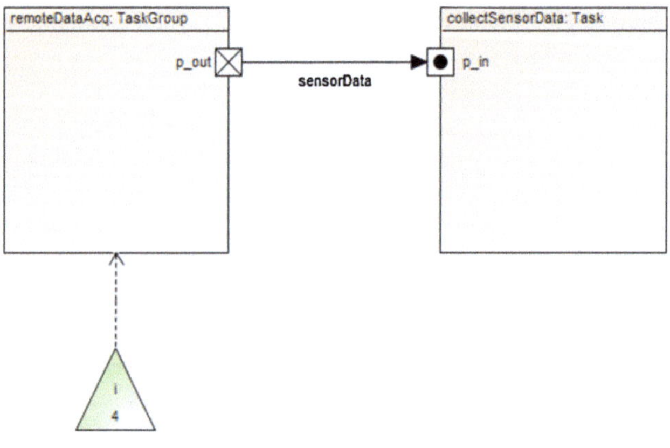

Fig. 1.1 Task model for a simple application

the mapping aspect. For more details about the semantics of replicators and how connections to replicated objects are handled, see [9].

When we zoom in on task group 'remoteDataAcq', we discover that it consists of two tasks: task 'dataAcq' and task 'convertData'. This is shown in Fig. 1.2. Note that port 'p3' of task 'convertData' is associated with port p_{out} of the task group. Port associations are necessary to specify how the task group objects are embedded in the higher level hierarchical task model.

Figure 1.3 shows the specification of task 'dataAcq' in the form of an algorithm model. The algorithm model specifies how the data is obtained, in this case simply an API function call of one of the hardware components of the node to which the task will be mapped. Note, that timing issues (e.g. sampling rate of the ADC) are omitted in this example.

Tasks can have resource requirements that may limit the possible mappings to physical components (nodes). A task can have a specified memory footprint, a computation load (nrIOPs and nrFLOps), and a set of required hardware resources (a list of device type names). The memory footprint value is used to check if a node can accommodate that task in its memory. The computational load is used to derive overall calculation time of that task on a node. If a behavioral model is associated with the task, the computational loads specified in the behavioral model are used instead. Finally, a task can only be mapped to a node that provides the devices that are listed in the required hardware resources attribute.

Before a task model can be simulated, the different modeling operators (like hierarchy, replication and parameterization) must first be resolved.

The specification of a task model is organized as a collection of task diagrams. Each task diagram specifies exactly one task group. The task diagram that specifies the task group at the top of the task hierarchy is called the context diagram. This diagram defines the system boundary of the system that is modeled as well as its relations with the context.

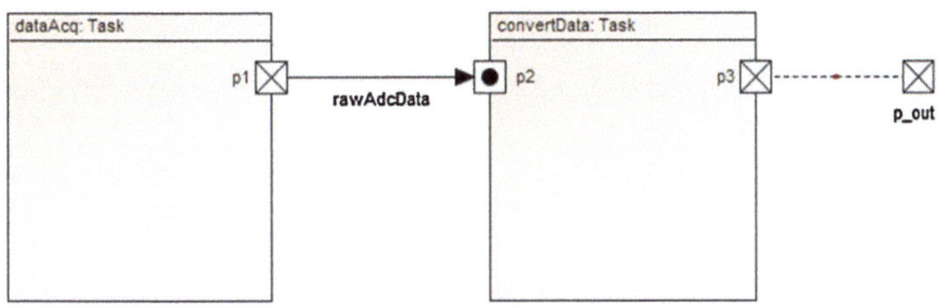

Fig. 1.2 Decomposition of task group remoteDataAcq

Fig. 1.3 Algorithmic specification of task dataAcq

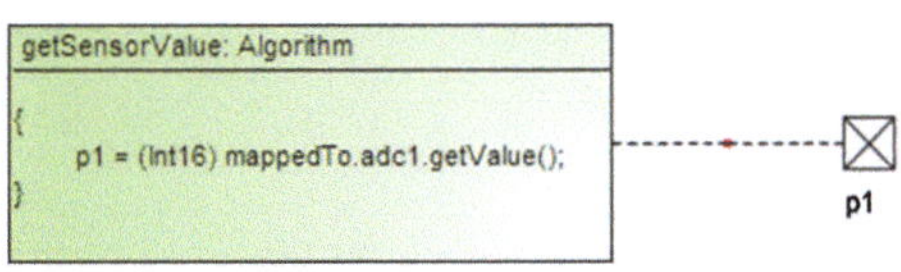

1.2.5 The Behavioral Aspect

The behavioral aspect is covered by a behavioral model. A behavioral model is used to specify the functional behaviour of a task. For this purpose, it uses similar semantics as a UML activity diagram [10]. However, the behavioral model does not support fork()/join() and barrier constructs, as these constructs are associated with modeling (dynamic) concurrency and covered by the task model. The behavioral model only captures purely sequential behaviour inside a task.

A behavioral model consists of exactly one behavioral diagram. A behavioral diagram specifies a sequence of operations. A behavioral model must have exactly one entry point and one or more exit points. Control flows specify the sequencing of the operations. A control flow is always associated with two (not necessarily distinct) operations: a source operation and a destination operation. Hence, a control flow has a direction. Operations must have one or more control flows leaving that operation and zero or more control flows entering that operation. Control flows may have conditional expressions involving data variables associated with them. Whenever an operation is finished, the control flow whose condition evaluates 'true' is taken to determine the next operation to execute. It is the responsibility of the system modeler that exactly one of the control flows leaving an operation can be taken if the operation finishes. If none or more than one of the leaving control flows can be 'taken', an error will result during simulation.

Continuing on the model of our simple application, Fig. 1.4 shows the behavioral specification associated with task 'convertData'.

The behavioral aspect supports different types of operations. Supported operation types are the processing operation, the communication operation (either send or receive) and the delay operation.

A processing operation performs a certain calculation involving a specified number of integers and floating point instructions.

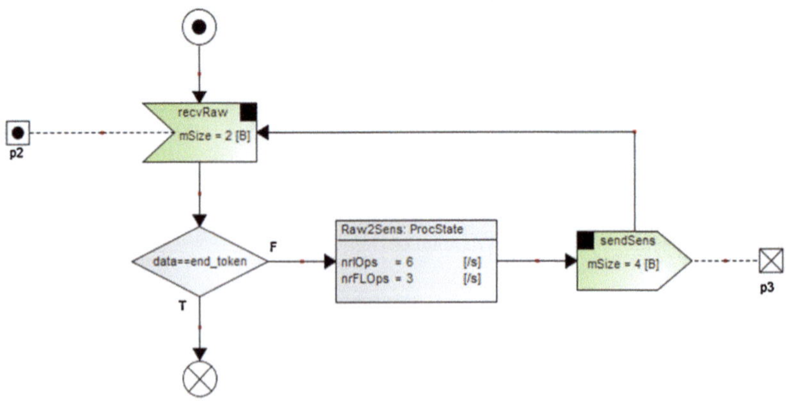

Fig. 1.4 Behavioral specification for task convertData

A communication operation sends or receives a specified number of bytes from a Task Port. Each communication operation specifies a synchronization kind. This can be 'blocked', 'non-blocked' or 'timeout'. Blocked communication waits (blocks) indefinitely until the communication is completed. Non-blocked communication either completes the communication without any need to wait (data message is already in the input port's queue), or continues the execution without doing any communication at all. Timeout communication waits at most the specified timeout value, and communicates the complete message within the specified timeout interval, or continues after the timeout without having communicated any data at all.

A delay operation simply waits (pauses) the specified amount of time.

In Fig. 1.4, both the send as well as the receive operations use blocked semantics, indicated by the black filled rectangles in the corners of the communication operation's symbols. For more details about the graphical syntax and exact operational semantics of a behavioral specification, see [9].

1.2.6 The Physical Aspect

The physical aspect is covered by a physical model. The physical model describes an abstraction of the physical resources of the system being modeled. It models the hardware resources that are used to implement/run an application Hardware resources include processing resources (processors, cores), communication resources (communication interfaces, communication networks), storage resources (memory) and energy resources (power supply, battery). Hardware resources can be shared (processor, memory, network) or can be consumed and replenished (energy).

When a tasks and task groups are mapped to hardware resources (the nodes), the implied semantics is that these hardware resources must be shared among the mapped tasks. Node resources are shared by mapped tasks and channels are shared by the data flows between tasks mapped to neighboring nodes. The time-sharing policy can be specified using a scheduling or media access arbitration algorithm.

When the available amount of resources, the number of competing parties as well as the provided scheduling policies are all specified, one can, through simulation, derive the timeliness of the behaviour of the system as well as other emergent behavioral characteristics. With respect to memory and energy usage, one can verify if the available resources are sufficient to drive the system for a specified time span.

A physical model describes the hardware resources of the system in terms of nodes that are interconnected in a network. The nodes themselves are defined in a node definition, in which all relevant hardware components are specified, together with their properties. For example, a node will typically have a processing component with properties such as processing capacity (nrIOps/core/s, nrFLOps/core/s) and power requirements; a memory component with properties data storage capacity (memory size); an optional battery component with property energy storage capacity; and a communication hardware component with properties data transfer capacity (bandwidth) and power requirements.

All hardware resources may have more than one operational mode, in which case each mode has an associated power usage and service capacity.

The physical model of our simple example application is given in Fig. 1.5. This model consists of two nodes (shown as gray colored rectangular shapes), connected via their communication interfaces (shown as small orange colored rectangles) to a communication channel (the orange colored rectangular shape). One of the nodes (myHost) is of type 'PC', while the other node ('nd') is of type 'GNode'. The latter node is acted upon by replicator 'j'.

The hardware resources and their properties associated with a certain type of node is specified in a node definition. Figure 1.6 shows the graphical node definition for a GNode.

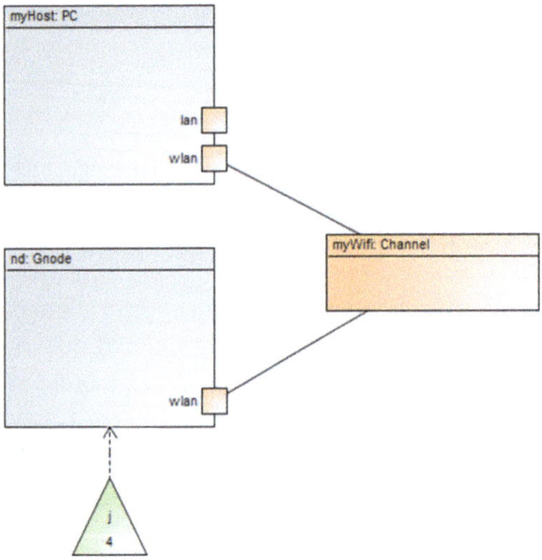

Fig. 1.5 Physical model for example application

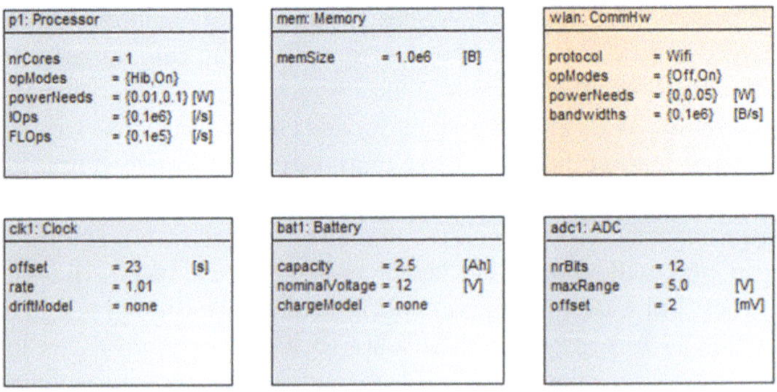

Fig. 1.6 Node definition for a Gnode

1.2.7 The Mapping Aspect

The mapping aspect is covered by a mapping model. A mapping model specifies how the tasks and other elements of a task model are assigned to the hardware resources of the physical model: the assignment means that the particular resource is involved in or responsible for the implementation of the task's functionalities. It should be emphasized that multiple components of the task model can be assigned to the same resource, i.e. functional components "compete" for shared resources; consequently resource access may need synchronization and scheduling. More specifically, tasks (and task groups) are assigned to nodes and the ports are assigned to the communication interfaces of the nodes. The mapping of the dataflow links to channels is not made explicit: this mapping is implied by the port mapping. The mapping specifications can be distributed over one or several mapping models. All diagrams together must specify a complete mapping, which must be both valid and consistent.

Figure 1.7 shows the mapping specification for the example introduced earlier. The mapping specification contains the task and task groups of the task model and the nodes of the physical model. Two kinds of mappings can be specified: task to node mappings and port to communication interface mappings. The task to node mappings are shown as dashed arrows, with a yellow background. The port to communication interface mappings are just dashed arrows. The mapping specification is completed by providing a replicator expression on the task to node mapping for the remoteDataAcq task group object. This replicator expression specifies which instance of the replicated

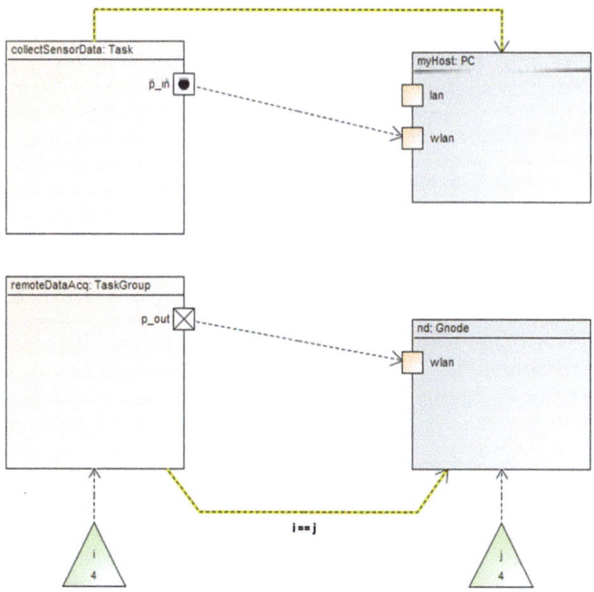

Fig. 1.7 Mapping specification for the example application

task group is to be mapped to what instance of the replicated node. Finally, the modeling language supports the specification of parameters in all system aspects. Parameters were not used in our example application. See [9] for more information.

1.2.8 Conclusions

In the previous sections, the DEMANES modeling language was introduced. Four modeling aspects have been detailed: the task, behavioral, physical and mapping aspect. The DEMANES modeling language will allow system designers to model large scale networked systems using a graphical representation. This graphical modeling language has been implemented using a commercial meta-modeling tool, as will be explained in Sect. 3.2. A set of constraint rules have been implemented using this tool, allowing designers to easily check their models for missing or inconsistent specifications.

The next section will elaborate on how the DEMANES modeling language can be used to quantify certain properties of the design. These properties include (among others) timeliness of system behaviour, throughput, resource use and resource efficiency.

The language described up to this point is comparable to other system modeling languages. Section 1.4 will explain how the DEMANES language can capture the reconfigurability aspect of systems. For this purpose, a new language construct will be introduced.

1.3 Model-Based Derivation of Key Performance Indicators

The modeling framework introduced provides rich capabilities to describe the system-under-design from different aspects, such as functionalities, hardware configuration, communication. The rigorous modeling allows for specifying the design, for communicating and documenting design choices but this is merely the first step. Our main goal is to characterize the design alternatives quantitatively in order to guide the designer along the design process: ideally, the design alternatives should be characterized in such a way that the derived properties should directly be comparable to key performance indicators.

Models used during the systems engineering process serve different purposes ranging from communication (among users and designers), documentation to evaluation, building and maintenance [11]. This section focuses on the use of models for design evaluation: along the design process the system designer has to make informed decisions when selecting the most "promising" design alternative. The selection should be driven by quantified properties of the design. These properties are originated in the design of components, compositions, parameters; and in the execution scenarios, i.e. the interactions between the systems designed and its embedding environment.

The model-based engineering approach formalizes all relevant aspects of the design in models and thus gives the formal foundation for deriving the emerging properties of the design. Frequently, the quantified design properties are aggregated in a "design quality measure" and used to guide a constrained design optimization process. The model based derivation of the design properties is just a manifestation of old and established engineering approach, namely use models to predict system behaviour [12].

The model based derivation of design properties and its use in "evolving" the system go beyond strictly design-time activities [13]. Generally speaking, the driving forces behind system evolution are "keeping operational" or "making it better" the system implemented as expressed in a quality measure. In runtime reconfigurable designs the calculation of the emerging system properties is carried out during the nominal operation of the systems to detect anomalies and consequently initiate and guide redesign (optimization) in runtime. Due to the possibly prohibitively large design space and the complexity of the design process the scope of the runtime redesign (i.e. the monitored set of key performance indicators and the investigated design alternatives) should be constrained [14]. For further details about the runtime–design time trade-off see Sect. 2.2.

1.3.1 Deriving the Key Performance Indicators

The model of a system is built from components defining particular elements of the design, e.g. tasks (functionalities), processors, communication interfaces, etc. All these components are annotated by attributes defining properties relevant for the implementation. Unfortunately these attributes in themselves do not say too much about the quality of the system as a whole. The attributes reflect very low level properties, which cannot directly be put side by side with the requirements set on system level. For example tasks are characterized by (among others) their computation demand (e.g. the number of floating point operations to be executed per invocation), but they do not determine directly the response time to the triggering event. The response time is an integral characteristic influenced by the computation demands, the task dependencies, the task allocation, scheduling, hardware characteristics, etc. The models describing the design should be made executable, where the execution is defined as contribution to the calculation of system level characteristics. The subset of the (emerging) system level characteristics relevant to the design (i.e. the related application) at hand is called *key performance indicators* (KPIs).

Figure 1.8 shows the process of the model-based support for system development. From the model execution, system level characteristics should be derived and compared to the requirements. If (some of) the requirements are not satisfied, the design should be modified. This is indicated as feedback in the figure. The modifications may target different aspects of the design, and accordingly adjustments in different models should be made. After the adjustments the KPI are recalculated (model execution) and design iteration starts.

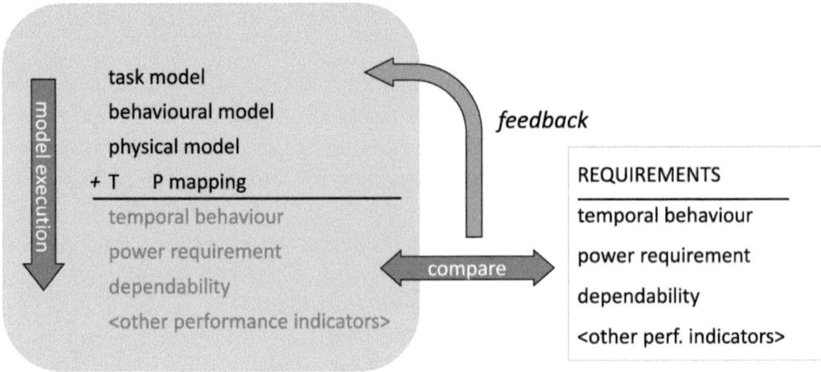

Fig. 1.8 Model-based system design

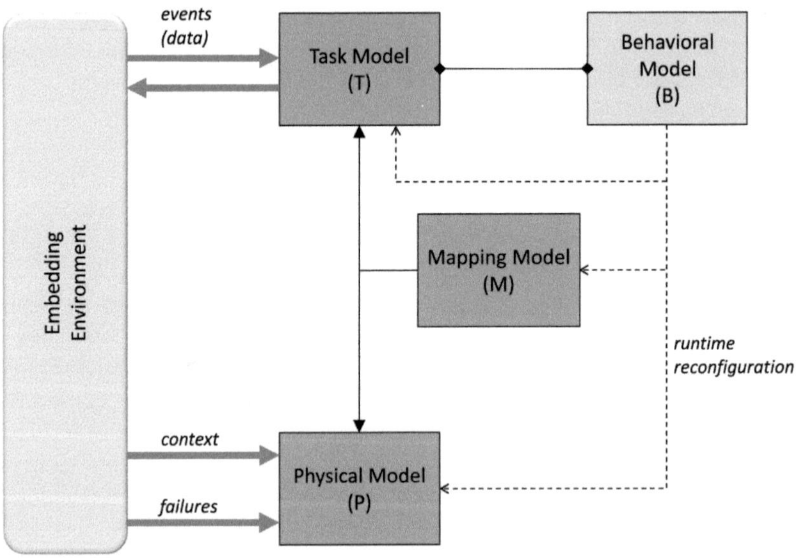

Fig. 1.9 Model execution in context

In order to derive the KPIs, the design should be placed into context: the model execution should be driven by event sequences resembling typical and/or critical real-world scenarios (see Fig. 1.9).

The events and other influences are generated by the embedding environment. The embedding environment constitutes for everything, which is beyond the border of the system to be designed.[2] The systems functionality (defined by the task models

[2]Obviously, the everything, which is beyond is not the whole world. Only those elements (incl. humans, eventually) are to be considered, which are connected to/influenced by the system to be

and the related behavioral models) is driven by the incoming events, such as new sensory data are available, user interface events, etc.

The embedding environment may influence the execution via changing the physical context of the system to be designed. Typical context changes are repositioning nodes with wireless communication links (i.e. mobile configurations), occurrence of obstacles, etc. These changes influence the task interactions indirectly via the changes in the physical configuration and/or its parameters, such as communication time or reachability. This results in different system level behaviors.

Also, system components may fail, which may have dramatic effect on performance or on the availability of certain functionalities. The figure indicates failures as external influences. This is merely a visualization of the approach that failure models are not considered as part of the modeling framework. The modeling framework can handle dynamics in the physical model but the causes of these dynamics is outside of the modeling framework. Failures are merely inserted to the execution, i.e. the creation of the failure is the responsibility of the embedding environment.

The dashed arrows in Fig. 1.9 indicate the impact of the runtime reconfiguration in modeling and model execution. Runtime reconfiguration needs functionalities built into the system, which monitor the execution and system health conditions and if needed makes adjustments in the system itself. The reconfiguration functionalities are reflected as dedicated tasks and attached behavioral models (with their own resource demand). The result of the reconfiguration is actuation on tasks (e.g. changing algorithms or parameters), on physical resources (e.g. adjusting transmission power, processor clock frequency, etc.) or on the mapping (e.g. reallocating tasks, rerouting communication). In the model-based design approach the reconfiguration results in changes in the models describing the system itself to reflect the changes in the real implementation. If the reconfiguration is successful the deterioration of the KPIs is less than in static designs. The runtime reconfiguration capability does not influence the calculation of the KPIs.[3]

The KPIs are functions of the relation between resource demands of the functionalities and the actually available resources, which may (and usually does) depend on conditions defined by the embedding environment. Figure 1.10 shows the flow of deriving the KPIs. The task model (T) and the associated behavioral model (B) driven by the event sequence (generated in the embedding environment) define the demands. From the annotated models and the events received the instantaneous computation demand ($d.cp$), communication demand (d.cm) and auxiliary resource (e.g. special peripheral device) demand ($d.r$) can be calculated. On the supply side the actual availability ($s.cp$, $s.cm$, $s.r$, respectively) is determined. For example effective bandwidth on a given communication link depends on the distance between the

(Footnote 2 continued)
designed. Usually these relevant elements are identified in the use-case models. Use case models are not considered here, they are assumed to be well-defined and stable.

[3]Deriving KPIs for runtime reconfigurable systems requires evaluation tools allowing model changes during the evaluation cycle. This is merely a tool implementation issue and will be detailed in Chap. 3. Note that the KPI calculation processes should be part of the implemented system itself to guide the runtime reconfiguration.

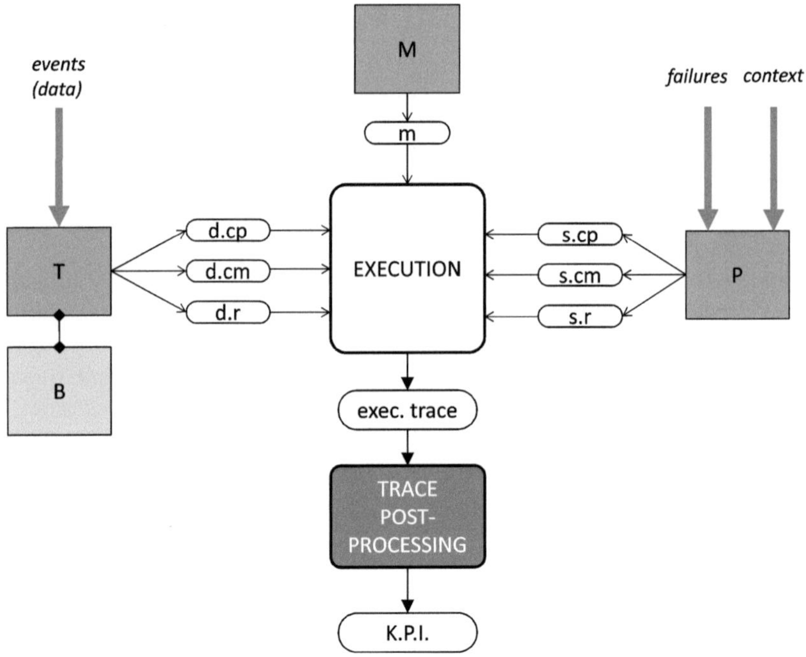

Fig. 1.10 The process of deriving KPIs (runtime reconfiguration related information flow is not shown)

communicating nodes, the obstacles in the physical environment, the transmit power setting, the antenna gains, etc. The physical model (P) should define all necessary component models and parameters to calculate the instantaneous supply.[4]

The correct execution of tasks (based on the models and with the associated behaviors) means satisfying the instantaneous demands by using the instantaneous supplies. As mentioned earlier the schedule of the tasks is determined by the external events. Due to the limited availability of the (shared) supply, the tasks compete for the resources thus task executions influence each other. This operation can be naturally mapped into discrete event simulation semantics: state changes in the system are caused only by external events (i.e. all influences from the embedding environment, represented by the gray arrows) or internal events generated by the tasks and the

[4]The underlying models for determining the instantaneous supply are typically very complex and the construction of these models go beyond the competence of the system designer (e.g. deriving channel models for wireless communication). Consequently the system designers work should be supported with parameterizable model libraries. In this case the designer just has to identify the matching model classes and has to set the parameters according to the scenario to be investigated. Many times determining the instantaneous supply is a computationally demanding process. The system designer has to find the balance between the fidelity and complexity.

components of the physical model.[5] Between events the system state remains constant. Consequently the execution of the model is not time but event driven: between the (time-stamped) events nothing happens, there is no need for continuous time progression. Execution time jumps from one event to the next one.

The result of the execution is the execution trace. The execution trace is the time-stamped list of events associated with the state of the model components.[6] The execution trace is the complete history of the execution of a particular scenario. As the execution trace the complete history including full system state, KPIs can be calculated off-line from the trace. The actual definition of the KPIs is case (and eventually scenario) specific and thus it is advantageous to keep the KPI calculation separate from the execution of scenarios.[7] It should be noted that in case of runtime reconfiguration certain KPIs are calculated by the task(s) responsible for controlling the reconfiguration.[8] This is a runtime activity defined by the behavioral model and it is not shown explicitly in the figure. The figure covers only the KPI calculation process for the evaluation of the system design.

1.3.2 Conclusions

System designers must make informed decisions when selecting the most "promising" design alternative. In order to assess the various system designs, KPIs are introduced. KPIs are system level characteristics are directly related to system requirements and relevant to the design.

The evaluation of a system design is executed by placing the design in its context. Typical real-world scenarios are simulated and the generated event sequences, operational mode changes are fed into the system model (e.g. a system failure event that can be injected into the system during an evaluation—and calculate how the performance of the system deteriorates). The resulting KPIs are evaluated in the domain of system requirements. If the experienced behavior is incorrect (i.e. certain requirements are not satisfied) adjustments should be made on the design (or the requirements should

[5]In reality computing nodes run schedulers to control access to the processor and other physical resources. The scheduler is typically a part of the runtime environment (operating system) managing the nodes operation. The proper execution of the system model requires the model of the scheduler also because the scheduler has the primary control on the local (in-node) resource access. The model of the scheduler is used by the EXECUTION block of the Fig. 1.10.

[6]In order to preserve memory in practical implementations only the state changes are stored (which is a much smaller set than the full system state as typically only a few components change states in response to an event). Conceptually it is the same as listing the complete system state. For the sake of simplicity we assume direct access to the full system state.

[7]The system design evaluation process should be supported by tools providing unified (standard) execution trace representation and post-processing libraries for filtering and calculating frequently used KPIs (e.g. energy consumption of components, utilization of resources, availability of functionalities, etc.). See Chap. 3 for details.

[8]For details about runtime reconfiguration solutions see Sects. 2.3 and 2.4.

be renegotiated). The design can be modified in design time (i.e. the designer goes back to the "drawing board", makes the necessary modifications (based on his/her experience, insight)) or in runtime.

The runtime "redesign" allows for flexibility and robustness on levels hardly achievable with static designs—for more details see the upcoming chapters of the book. A straight consequence of the usage of the runtime redesign capability is that the KPIs (which guide the redesign process) for the different design alternatives (created by the automatic reconfiguration process relying on built-in "design knowledge") should be derived during the operation of the system (i.e. in real-time). It should be emphasized that the runtime derivation the KPIs requires the system model to be available in the deployed version of the system, i.e. a "self-reflection" capability should be provided by the implementation [15].

1.4 Modeling of Runtime Reconfiguration

In a continuously changing environment, the static design of an networked embedded system (NES) will fall short if the deployed system has to operate over long periods of time. Ultimately, deterioration come to the system as a result of failing hardware, changed operating conditions or unexpected changes in the targeted system application. In order to retain robust and reliable performance, it seems essential to allow for some form of adaptivity in the system design. This adaptivity will have to be able to change the *modus operandi* of the system at runtime, i.e. it will have to be able to change the configuration of the NES.

In this section, a system architecture is described capable of incorporating runtime reconfiguration capabilities in a structured way. The goal of this architecture is to realize a reconfiguration functionality in a distributed and service-oriented manner, in order to improve dependability and scalability. The main advantage of the proposed architecture is that it has minimal impact on the primary data processing path, i.e. the design of the primary function and the reconfiguration function can be kept independent to a great extent. As a consequence, the verifiability of the primary data processing function can be maintained, as different system configurations are adopted. For more detailed discussion, see [16].

Earlier studies of runtime reconfiguration often use finite state machines in which states denote a configuration and the transitions specify a reconfiguration action. An example of this approach is shown in [17] in which not only the designed system, but also the reconfiguration process itself is modeled as a finite state machine.

The ARCM (Architecture Runtime Configuration Management) [18] makes explicit use of this state machine approach of modeling runtime reconfiguration by maintaining information about the used configurations, and representing the reconfiguration process posterior to running the system. By doing so, the reconfiguration process is made more insightful, and increases the understandability and visibility of the adaptivity of the system.

The finite state machine approach addresses the problem of the reconfiguration process, but not the "reasoning" guiding the system reconfiguration. In other words, system changes reflect decisions made during design time and often triggered by well known events. In general, there is no runtime design space exploration to determine the system state to which the system should transit.

In [19], the concept of meta-cognition is discussed. The authors propose that a meta-cognitive loop should be added to an existing system, which observes, reasons and adapts the system if a failure or poor performance occurs. This concept also underlies the ideas of the MAPE-K [20] architecture. An example of meta-cognitive loop is the ReCoNet definition [21]. In this definition, a system is formally defined as a combination of a *task graph*, an *architecture graph*, a *task binding* and a *communication binding*. By changing task and/or communication bindings, the modeled system changes its meaning. Using these primitives, phenomena such as dynamic routing, task migration, hardware/software morphing and on-line partitioning can be modeled.

In the rest of this section, we introduce a methodology to describe runtime reconfigurable systems based on the combination of design patterns for reconfiguration and a formal methods to describe adaptive systems.

1.4.1 Model Based Design for Reconfiguration

Fig. 1.11 shows a functional scheme of the architecture by which a runtime reconfigurable system can be designed. The main structure of the reconfiguration architecture is visible in this figure as two separated functional layers: a primary data path (at the top) and a "management" layer (at the bottom).

The primary functionality of the system is realized by a task network, via the invocation of a chain of associated functional primitives. This task network is built during initialization time according to a design-time specification. Reconfiguration tasks are set to run in the management layer, where they can influence the primary

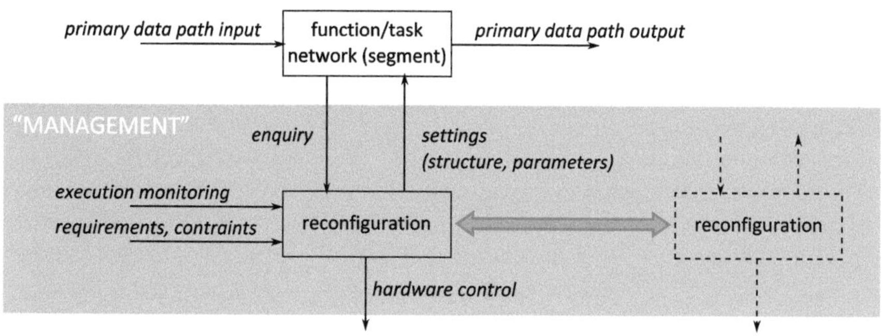

Fig. 1.11 The reconfiguration of the primary data path

path in case of need. The execution status of the system, such as the quality of the generated results, the availability/performance of the hardware components, or the characteristics of communication links, form the basis on which the reconfiguration functionalities makes its decisions. The reconfiguration layer may act both on the "software side" of the configuration (e.g. by selecting a different algorithm for implementing a task) and on the "hardware side" of the configuration such as suspending or awaking components or changing transmission power. An advantage of this architecture is that the primary path is coherent and analyzable for all states of the reconfiguration.

The challenge faced by the management layer is to realize the required functional properties with a limited set of resources, while satisfying non-functional requirements such as timely responses, dependability, power consumption etc. Such a challenge is currently mostly faced during design time and by a system designer. One successful method for dealing with this challenge is the model-based system design methodology proposed in [22]. It should be emphasized that the implementation of the management functionalities in runtime are carried out by (dedicated) tasks, forming a task network. Such extra tasks should be incorporated in the task model of the complete systems, i.e. their resource needs can be taken into account during design and execution.

In order to reason about the reconfiguration of a system in runtime, the system should explicitly represent its architecture, its components and their properties in the runtime version. This (perhaps partial) representation is used when the system reasons about its own reconfiguration. The methods for representation and reasoning about reconfiguration will be discussed in more detail in Sect. 2.4.

1.4.2 Reconfiguration Types and Basic Architectures

Traditionally the system architecture design process is completed off-line, i.e. during design-time. This means that it is mostly done before any implementation takes place, and completely done before deployment of the system itself. It is often assumed that the task model, the available hardware resources, and their characteristics are known at this time and uncertainties are assumed to be low. These assumptions are however overly optimistic for large-scale networked embedded systems, as in many cases they are deployed in unknown and sometimes "hostile" environments. Under these circumstances system components may fail, and changing operational conditions are common.

In order to realize a network of reconfigurable components, tasks in the primary data path must implement an interface that allow starting or stopping, reconnecting or parameterizing them. It is the responsibility of the management tasks to activate these interfaces and ensure that all dependencies of the functional services are fulfilled. Also, the parameters used by all the primitives should be configured by the management layer, for which a system for setting and retrieving parameters is employed.

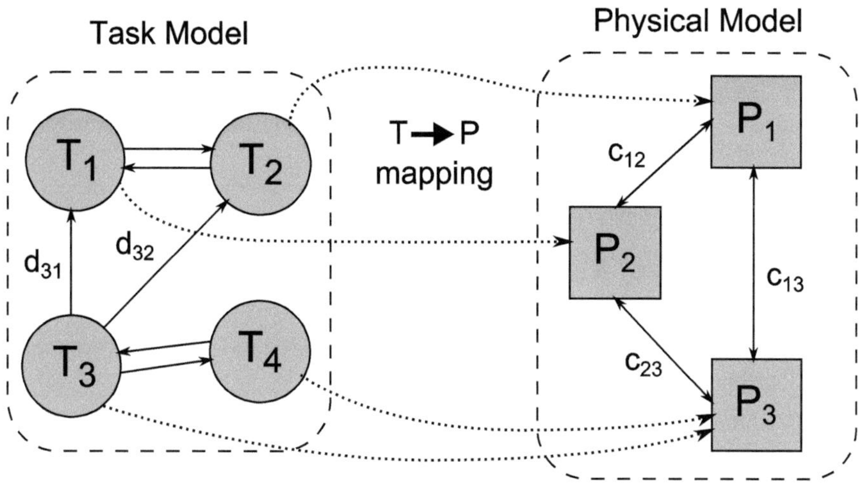

Fig. 1.12 Task and physical model. The *squares* marked T_x denote tasks, connected by task dependencies, correspondingly the *circles* marked P_x denote processors connected by communication links

The runtime reconfiguration is analogous to altering the task graph or re-mapping the tasks to the physical (hardware) components.

Using the task model and the physical model (for example the one in Fig. 1.12), the reconfiguration problem can now be described as a constraint optimization problem for assigning the tasks T_q to processors P_i (i.e. assigning the tasks to nodes), in which the available memory and execution time requirements define the constraints to be satisfied. Furthermore, the data exchange paths between tasks have to be associated with communication links c_{ij} between nodes. These communication links also have limited resources (i.e. bandwidth). The objective of the optimization is set by the designer in the form of objective function (or performance criteria) and typically includes some combination of KPIs (e.g. a measure of the primary functions' performance, system throughput, system lifetime (for battery powered solutions)).

The following characteristics of the proposed scheme should be emphasized:

- At any time instant, the function/task network is a snapshot of the possible variants and mappings. The alternatives may not be explicitly enumerated but can be the result of a reasoning (problem solving) process.
- The scheme explicitly supports the separation-of-concerns principle. The reconfiguration mechanism should be designed and implemented relatively independently.
- The reconfiguration can be a resource demanding activity. The scheme allows for tuning the "intelligence level" of the reconfiguration depending on the performance of the hardware configuration—virtually leaving the signal processing aspect uninfluenced.
- The scheme is applicable to reconfigurations on "various levels of granularity": task, node and system levels. This means the reconfiguration scales from fine grade

distributed to centralized. Needless to say, distributed reconfiguration may need cooperation among the reconfiguration functionalities.
- From execution point of view, the reconfiguration functionalities should be included in the task graph as one or more cooperating tasks and their resource demand should be accounted for.

1.4.3 Modeling of Runtime Reconfigurable NESs

In order to model reconfigurable NESs we need to extend the modeling framework as proposed in Sect. 1.2.

The role of the models in runtime
The is a process that changes the system during its execution. When considered at the system design phase, reconfiguration reasons on the system design model to facilitate the evaluation of the performance and resource demand of the reconfiguration. After the system deployment phase, the runtime reconfiguration acts on the real hardware and/or on the software components comprising the system as well as on the (partial) system model serving as the representation of the current configuration in runtime (in order to keep the representation of the configuration up-to-date). It should be emphasized that the system model is an explicit part of the implementation, that is, it is stored and used during the nominal operation. Such model is used to recalculate KPIs and to reason about the configuration changes. Reconfiguration functionalities can be supported by middleware services, which assuresw unified access to the components. Middleware services will be considered in more details in Chap. 3.

In order to model the reconfiguration capabilities, we introduce modeling primitives expressing particular model transformations, i.e. indicating the "actuation points" where changes in the system model are allowed. For the "scope of actuation" on the system we use the following configuration types (categorized by models):

Re-parameterization: Changing a parameter of an existing component

1. Task model: Changing task (functional) parameters
2. Physical model: Modification of operational modes, hardware parameters, clock, etc.

Re-instantiation: Creating or destroying components

1. Task model: Changing the execution state of the tasks, creating/destroying instances of tasks
2. Physical model: Adding or removing nodes and other (hardware) resources

Rewiring: Creating or destroying connections between components

1. Task model: Changing the connectivity between tasks (by changing the "connection matrix")
2. Physical model: Changing connectivity between nodes (adding/removing of channels or changing channel parameters)

Relocation: Migrating functionalities from one component to another

1. Mapping model: Allocation/deallocation/reallocation of tasks and/or changing the link-to-channel mapping

The extension of the modeling framework
The reconfiguration interactions among the model components are represented by handles. The handle indicates both the modeling object to be influenced as well as the task component initiating the reconfiguration actions. This corresponds to the Acceptor handle (depicted as a stick with a semicircle) and the Issuer handle (a stick with a black dot) respectively. A simple example is shown in Fig. 1.13 where the Issuer handle 'h_reconf' is associated with the Task 'reconfigurationTask'. In this example, the reconfiguration action can change the parameters of the set of Tasks of type 'ApplicationA' via the Acceptor handle 'modifyPars'.

Depending on which modeling primitive an Acceptor handle is attached to, the reconfiguration action depicts another type of reconfiguration. Figure 1.13 shows the reconfiguration of a Task (reconfiguration type 1), the re-parametrization of a sub component. This type of reconfiguration is bound to a parameter in the model, and an annotation associated with the reconfiguration mapping can indicate what parameter exactly can be reconfigured.

Alternatively the Acceptor handle could be attached to a replicator. This indicates that during runtime, the reconfiguration task can alter how many instances of a specific task exists. In practice this means that the reconfiguration task can either kill an existing task, or it can start a new one. This is reconfiguration type 2. However, instantiation or deletion of new nodes (type 2b) cannot be initialized from within a task, but is actually an event which occurs during runtime. To create a model that allows for a certain type of nodes entering the system, a replicator must be included in the model, replicating the corresponding type of node. The cardinality specifier of this replicator is however not determined at design time, so it gets a value '*'.

For the third type of reconfiguration, the Acceptor handle might be attached to a connection between elements. In this type of reconfiguration the wiring between

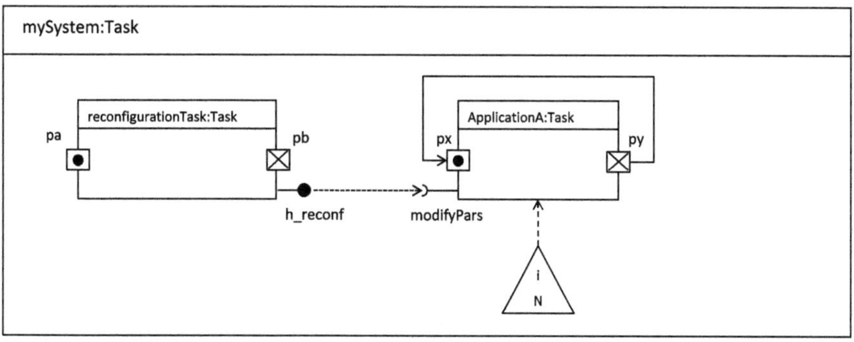

Fig. 1.13 A graphical model representing the reconfiguration of a parameter

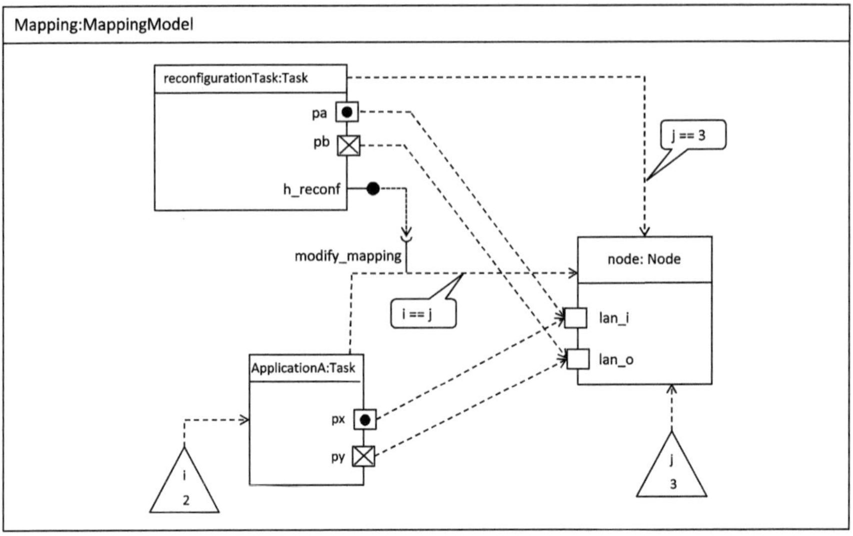

Fig. 1.14 A graphical model representing the possible relocation of tasks in different nodes

model Tasks or between Nodes can change. In the modeling language this corresponds to a change in the expression that defines the filter of the connections between the replicated objects. In the implementation this corresponds to changing the destinations of messaging of the communicating Tasks or Nodes, or altering the physical Channel specifications.

Finally reconfiguration type 4 is shown in Fig. 1.14. In this type of reconfiguration the mapping of tasks to nodes are changed. In the modeling language this again means the expression of the mapping filter changes, in a similar fashion as configuration type 3. It may seem superfluous considering reconfiguration type 2, by simply stopping a task in one node, and starting it in another. However in some cases a model may be required in which there is always a fixed number of a specific task. In those cases it is more intuitive to relocate a task rather than stopping it and starting a new instance.

Two peculiarities of the reconfiguration modeling should be noted:

1. Replicators may have a dynamic cardinality specifier ('*'). The associated component acts as a "placeholder" indicating that type of component may be created during runtime (an unspecified number of instances) and after all connections and mappings are made it becomes part of the system.
2. In order to provide a notation for the reconfigurable mapping, an acceptor-type handle can also be associated with a task-to-node or connection-to-channel mapping edge (as in Fig. 1.14). It reflects the reconfiguration scenario, which may change the mapping. Consequently in these cases the Mapping model should be interpreted as a snapshot of the (evolving) mapping configuration and at a particular time instance the actual mapping could be different. Typically Mapping

models show the initial configuration from which the operation of the system starts.

The reconfiguration primitives and the underlying reconfiguration scheme are very generic and have positive and negative characteristics. On the positive side it allows the system designer to express a wide variety of reconfiguration solutions. Practically, what the designer can do in design-time, now the same can be done in runtime too. On the negative side (due to the unconstrained "freedom") many ineffective and unsafe reconfiguration schemes can be created. Systems designers definitely need strong design guidelines (and supporting tools). A set of reconfiguration design patterns and guidelines shall be discussed in Sect. 2.3.

1.4.4 Conclusions

Concluding this section, we have described an architecture for runtime reconfigurable systems based on a two-layered approach. In these two layers there are the primary tasks in one layer, and the secondary "management" tasks in the other. By extending the modeling language with reconfiguration primitives, we can now describe a system that alters its behaviour in runtime, i.e. not all design decisions are made explicit beforehand. Using these primitives we can describe four basic types of reconfiguration: re-parametrization, re-instantiation, rewiring and relocation.

1.5 Conclusions

This chapter focuses on model-based engineering of large-scale complex dynamical systems. Distinguishing features of the targeted systems are serious resource constraints (e.g. processing capabilities, communication, energy supply), and the presence of demanding non-functional requirements, such as timeliness, robustness, lifetime and the capability of handling system evolution, i.e. runtime reconfiguration.

Existing modeling methodologies, languages and tools either focus on a single application domain, have limits on scalability or no capability to handle runtime reconfiguration.

A modeling methodology has been introduced, with special attention for modeling runtime reconfiguration aspects at design-time and scalability. The methodology is based on MAPE-K and uses feedback loops in the system model to model runtime reconfiguration. The modeling methodology is accompanied by a multi-aspect modeling language which allows system designers to model the system architecture from different aspects. The language includes concepts of cardinality, replication and runtime reconfiguration allowing modeling of large-scale systems in a concise and compact manner without loosing flexibility in the modeling process.

A scheme has been introduced for the derivation of system level characteristics, i.e. KPIs, which allow the evaluation of system designs and the selection of the most promising one.

References

1. J.O. Kephart, D.M. Chess, Computer **36**(1), 41 (2003)
2. ISO/IEC: International standard ISO/IEC 10746-3 (1996)
3. ISO/IEC: International standard ISO/IEC 10746-2 (1996)
4. ISO/IEC: International standard ISO/IEC 10746-4 (1998)
5. ISO/IEC: International standard ISO/IEC 10746-1 (1998)
6. Feiler, Gluch, in *Model-Based Engineering with AADL: An Introduction to the SAE Architecture Analysis and Design Language* (2012)
7. O.M. Group: OMG Systems modeling language (2012)
8. G. Karsai, F. Massacci, L. Osterweil, I. Schieferdecker, Computer **43**(5), 34 (2010)
9. T. Vogel, in *DEMANES* (2014)
10. O.M. Group: OMG Unified Modeling Language Superstructure Specification (2007)
11. B. Brown, Model-based systems engineering: revolution or evolution? (2011). http://www-01.ibm.com/common/ssi/cgi-bin/ssialias?infotype=SA&subtype=WH&htmlfid=RAW14289USEN
12. B. Morin, O. Barais, J.M. Jzquel, F. Fleurey, A. Solberg, IEEE Comput. 46–53 (2009). http://www.irisa.fr/triskell/publis/2009/Morin09f.pdf
13. G. Karsai, F. Massacci, L. Osterweil, I. Schieferdecker, Computer **43**(5), 34 (2010). doi:10.1109/MC.2010.135
14. T. Streichert, D. Koch, C. Haubelt, J. Teich, EURASIP J. Embed. Syst. **2006**(1), 042168 (2006). doi:10.1155/ES/2006/42168, http://jes.eurasipjournals.com/content/2006/1/042168
15. T. Gjerlufsen, M. Ingstrup, J. Olsen, Computer **42**(10), 61 (2009). doi:10.1109/MC.2009.325
16. C. van Leeuwen, J. Sijs, Z. Papp, in *Fusion* (2013)
17. J. Teich, M. Köster, in *Proceedings of the conference on Design, automation and test in Europe (IEEE Computer Society, 2002)*, pp. 559–566
18. J.C. Georgas, A. van der Hoek, R.N. Taylor, Computer **42**(10), 52 (2009)
19. M. Schmill, T. Oates, M. Anderson, S. Fults, D. Josyula, D. Perlis, S. Wilson, in *AAAI-08 Workshop on Metareasoning*, (Chicago, IL, 2008)
20. J. Kephart, D. Chess, Computer **36**(1), 41 (2003)
21. T. Streichert, D. Koch, C. Haubelt, J. Teich, EURASIP J. Embed. Syst. **2006**(1), 9 (2006)
22. G. Karsai, J. Sztipanovits, IEEE Intell. Syst. **14**(3), 46 (1999)

Chapter 2
Designing Reconfigurable Systems: Methodology and Guidelines

Zoltan Papp, Raul del Toro Matamoros, Coen van Leeuwen, Julio de Oliveira Filho, Andrei Pruteanu and Přemysl Šůcha

Abstract One of the major challenges when designing software for complex systems relates to a lack of a specific and comprehensive set of rules and methodologies. Even more so, adaptation to field conditions is difficult to model and implement on systems composed of a larger number of devices/components, such as distributed systems or systems of systems. On state-of-the-art technology such as wireless sensor/actuator networks and cyber-physical systems, addressing the lack of a compressive set of rules for their design and realization offers considerable benefits. If successfully realized, it can accelerate and simplify their design and implementation. The main contribution of this chapter is a clear set of rules that are specific for the design of adaptive networked embedded systems. To be more specific, we discuss design-time vs. runtime trade-offs, introduce design patterns for reconfigurable real-time monitoring and control, propose techniques for runtime design space exploration (managing runtime reconfiguration) and a systems engineering process for runtime reconfigurable systems. We provide guidelines for all stages of the architectural process and help system and software designers in choosing wisely specific algorithms and techniques. In conclusion, this chapter introduces a set of rules (methodologies) that are specific for designing adaptive networked embedded systems.

Z. Papp · C. van Leeuwen · J. de Oliveira Filho (✉)
TNO, Oude Waalsdorperweg 63, The Hague, The Netherlands
e-mail: julio.deoliveirafilho@tno.nl

C. van Leeuwen
e-mail: coen.vanleeuwen@tno.nl

R. del Toro Matamoros
UPM, Madrid, Spain
e-mail: raul.deltoro@car.upm-csic.es

A. Pruteanu
TU Delft, Postbus 5, Delft, The Netherlands
e-mail: a.s.pruteanu@tudelft.nl

P. Šůcha
Czech Technical University in Prague, Technická 2, 166 27 Prague 6, Czech Republic
e-mail: suchap@fel.cvut.cz

Z. Papp and G. Exarchakos (eds.), *Runtime Reconfiguration in Networked Embedded Systems*, Internet of Things, DOI 10.1007/978-981-10-0715-6_2

2.1 Introduction: Why Design for Runtime Reconfiguration?

System reconfiguration adapts the system to changes, and as such is carried out for satisfying a predefined goal. Finding the suitable system configuration, while maintaining all design and execution constraints is a demanding and knowledge intensive process. In essence, the runtime reconfiguration is carrying out system design activities during the operation of the system. Consequently all steps of the reconfiguration cycle assumes domain insight and generic engineering knowledge.

The reconfiguration schemes in DEMANES takes into consideration the particular challenges of the DEMANES application domains. These special challenges are:

- The primary functionality of the system should be kept in operation. For that, we advocate a separation-of-concerns approach where the primary datapath of the application and the reconfiguration functionality are clearly separated.
- The reconfiguration should be carried out under time constraints. It does not have to be necessarily fast, but it have to respect the response times that are expected to the system application.
- The typical underlying hardware is resource limited: computing capabilities, memory, communication capabilities, energy supply are severely constrained. Accordingly, related architectural and algorithmic challenges should be addressed.

2.1.1 Reasons for Reconfiguration

During the design process, the designer faces a number of questions about design alternatives and parameterization of the design. Implicitly or explicitly, every question is answered via evaluating alternatives, assigning a measure to them which reflects the "quality" of the alternative, and then selecting the "best" solution.

Except trivial cases the "question–answer" process is very complex: the number of design variables is typically large, there are many possible answers and there are interactions and dependencies among design decisions. The design variables define a design space, in which every design alternative is represented by a point, as depicted in Fig. 2.1. While exploring the design space, possibly complex and large number of requirements design constraints should be satisfied—and these obviously constrain the design space.

Furthermore when answering design questions, the designer makes assumptions, whose validity should be considered as preconditions for the validity and quality of the design. Figure 2.1a represents an (overly simplified) abstract design situation: the designed system is meant to monitor a connected process. There are three design dimensions and the designer have to select a particular combination of these three, that maximize the quality (Q) of the resulting system. All that, while satisfying all requirements and design constraints. Along such design, the designer makes certain

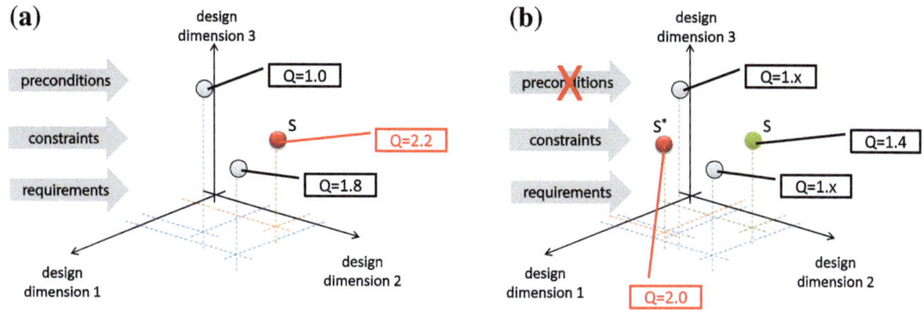

Fig. 2.1 Design space exploration

assumptions, such as the connected system is governed by linear dynamics; and/or there are some available resources to carry out certain calculations.

As an example, imagine that under the mentioned conditions, the design annotated with *S* is the best choice, and thus this combination of the free design parameters should be selected. If, for whatever reasons, the requirements or constraints change, or any of the design assumptions becomes invalid (Fig. 2.1b), then the design S may become sub-optimal (or even not feasible at all), and consequently the design should be adjusted. It means that the design space exploration should be continued to find a new solution via different selection of the design parameters (S^*).

When changes in requirements, constraints, or preconditions may happen during the nominal operation of the system then the design may be done such as to be prepared for the expected changes. For example, designing suboptimal system states that remain valid under the expected changes; or adding an "overlay mechanism" which isolates the system from these changes are perfectly valid options. The problem is that frequently such design time solutions for coping with changes result in complex, over-dimensioned designs–and the designed system remains "fragile" anyhow as it can only handle changes envisioned during the design.

An alternative, and in a number of aspects, more robust solution is implementing reconfigurationcapabilities for runtime. In this case the design space exploration and finding the most suitable design is a runtime activity. For that, the explicit representation of the design and some kind of "reasoning mechanism" should be part of the running implementation of the system. The design space exploration, design evaluation, and runtime decision making are resource demanding activities. In order to maintain the resource usage on an acceptable level and still assure temporal guaranties, the design space for the runtime design should be constrained. Thorough analysis (in design time) can be carried out to determine the scope of the runtime reconfiguration–carefully balancing between the resulting robustness and the required resources.

Runtime reconfiguration capabilities are typically used to address the following situations:

Hardware failure/degradation: This is typically results in loosing resources and assigned functionalities. The challenge is to find a different resource allocation, which can keep (at least) the critical functionalities running.

Communication failure/degradation: In distributed embedded systems the quality of the interconnection directly influences the correctness of the operation. Consider for example response deadlines which can be missed due to communication degradation. Changes in communication performance, or loosing the communication completely, may trigger reallocation of functionalities and/or switching to a different processing scheme, which has lower communication demand.

Energy considerations: In energy constrained implementations—for example, where certain subsystems run on battery power—assuring the required life-time of the system is of primary importance. If for reasons not accounted at design time, the energy level in some subsystems falls faster than expected, countermeasures can be introduced by, for example, restructuring and/or reallocating functionalities.

Changing user requirement: The life-cycle of deeply integrated monitoring and control systems is typically very long and meanwhile changes in user requirements are to be expected. On the other hand, these deeply integrated systems are continuously in use and new capabilities or modification of existing ones should be carried out on-the-fly, typically in a gradual fashion.

Changing operating conditions: As mentioned earlier, the system design builds on assumptions about the behavior of the coupled external processes. During operation these preconditions may become invalid and thus the system design should be adjusted to accommodate the situation.

Mobility, configuration changes: In certain application, the interconnection topology tends to change due to component mobility. As the interconnection topology has critical impact on the operation of functionalities implemented on the distributed platform, changes in topology should be managed. One alternative is to add functionalities and redundancy to maintain static topology, but often this leads to an overly expensive and even unfeasible solution. Instead, runtime reconfiguration capabilities may change processing architecture to "follow" the topological changes and assure functional robustness.

This list is not to meant to be exhaustive—instead it just indicates a few typical uses. It always should be kept in mind that managing reconfiguration is a resource demanding activity in itself. Building in runtime reconfiguration capabilities should always be justified. Consequently thorough analysis should be carried out to determine the scope of the reconfiguration–and sometimes solutions derived at the design phase may be the best approach. In the following, we discuss a bit more the trade-off between taking decisions at design time or at runtime.

2.2 The Design Time Versus Runtime Optimization Trade-Off

The main advantage of reconfiguration is ability to react to new situations at runtime. It provides flexibility to the system and it prepares it to situations that where not considered at design time. However, the reconfiguration itself has to often solve computationally complex problems, i.e. problems that cannot be optimally solved in polynomial time. Take for example the runtime re-mapping of tasks to the physical units described in Sect. 1.4.2. In practical applications, it is solved using heuristic methods in order to be able to react reasonably fast. By using heuristics, there is no guarantee how far from the optimal state the system will be after a reconfiguration is applied. That shows that the flexibility gained through reconfiguration often comes at cost of using a sub-optimal state.

Another aspect contributes to this sub-optimality. The essence of distributed systems is decentralization: it means that parameters and information about particular nodes is often spread out in the entire network and there is no a centralized authority having all the information necessary to perform the reconfiguration. Without a global view on the system state, it is often impossible to make decisions that lead the system to a global optimal state.

Despite the fact that the above described problems cannot be solved at the runtime efficiently, there are cases when design time optimization techniques can make the reconfiguration more efficient. There are no doubts that if something can be decided at the design time (with at least the same quality/efficiency as at the runtime) it should be decided there and it should not increase complexity of the reconfiguration. Moreover, at the design time there is more time to pre-compute, for example, several design patterns that can be applied at the runtime. The design time optimization cannot be generalized since it is a problem specific operation. However, we can categorize it into three classes:

Initial system state design: In applications such as wireless sensor networks, the reconfiguration cannot be powerful enough to guarantee optimal or near optimal system behaviour. The aim of the reconfiguration is often to keep the system alive. However, due to decentralized natural of the system or lack of time to perform some more sophisticated optimization, the key performance indicators can slowly get worse and worse. Even though this issue is very difficult to solve, the situation can be improved by a better initial solution. This solution can be found at the design time where a more complicated algorithm may be used to solve the problem. Consequently, the system at runtime starts with much better key performance indicators which may lead to better system behavior and also to its higher reliability.

System design optimization aimed at the reconfiguration: In many cases there is no possibility to carry over complexity of the reconfiguration from the runtime to the design time. However, it may be possible optimize properties of the system in order to make the reconfiguration more efficient, for example by determining the optimal number of nodes that guarantee sufficient reliability of the system.

Decomposition based design time optimization: Some applications allow decomposing the reconfiguration into decision that must be made at the runtime and the decisions that can be made at the design time. Typical examples are systems operating in a finite number of modes, for example, high/low battery capacity. Then the reconfiguration decides which mode will be selected at the given moment. In such a case the system designer can pre-compute patterns, for example a schedule of activities, to be applied in the system according the active mode. By doing so a part of the decision making is moved into the design time and can simplify the whole reconfiguration mechanism.

A good example of the first category is the runtime re-mapping of tasks to the nodes described in Sect. 1.4.2. The corresponding model is illustrated in Fig. 2.2. Task called 'reconfigurationTask' can stop a task ('Application') in one node ('Node') and start it in another one. This task mapping is controlled via'h_reconf' signal at the runtime. However, the initial mapping is computed at the design time and it is stored in 'reconfigurationTask' in property 'initial_mapping'.

In the first case, the objective of the design time optimization is the same as the objective of the reconfiguration. In the second category, the function at the design time optimization is different. Considering the same example of runtime re-mapping of tasks illustrated in Fig. 2.2 the system designer may be interested in such a physical network design that minimizes the number of needed nodes (m). Another aim of the design time optimization would be to determinate position of nodes (property

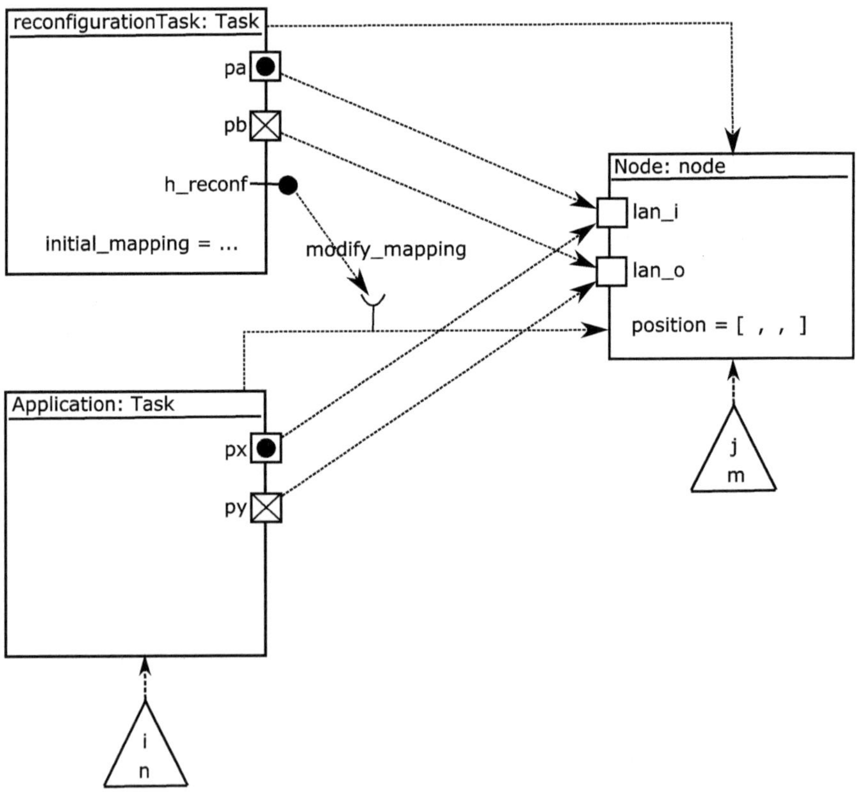

Fig. 2.2 Design time optimization

'position') such that each node has enough neighbors that can take over his tasks if needed. In this way we can guarantee certain reliability of the reconfiguration.

To illustrate the last category lets consider a slightly different scenario. The model in Fig. 2.3 considers tasks to be executed on nodes. In this case, the mapping between nodes and tasks is fixed. However, we consider two modes of the network, e.g. high battery and low battery state. If the network operates in high battery mode the network performs all the tasks while in the second one some tasks are deactivated. This is controlled by 'reconfigurationTask' via the replicator of 'Application' Task. For different modes there are different schedules of tasks stored in property 'schedule' in each Node. Since the task mapping is fixed the problem can be decomposed into two phases. The decision made at the runtime is responsible for selection of the network mode and the design time optimization assigns to each mode and each node a fixed schedule of activities.

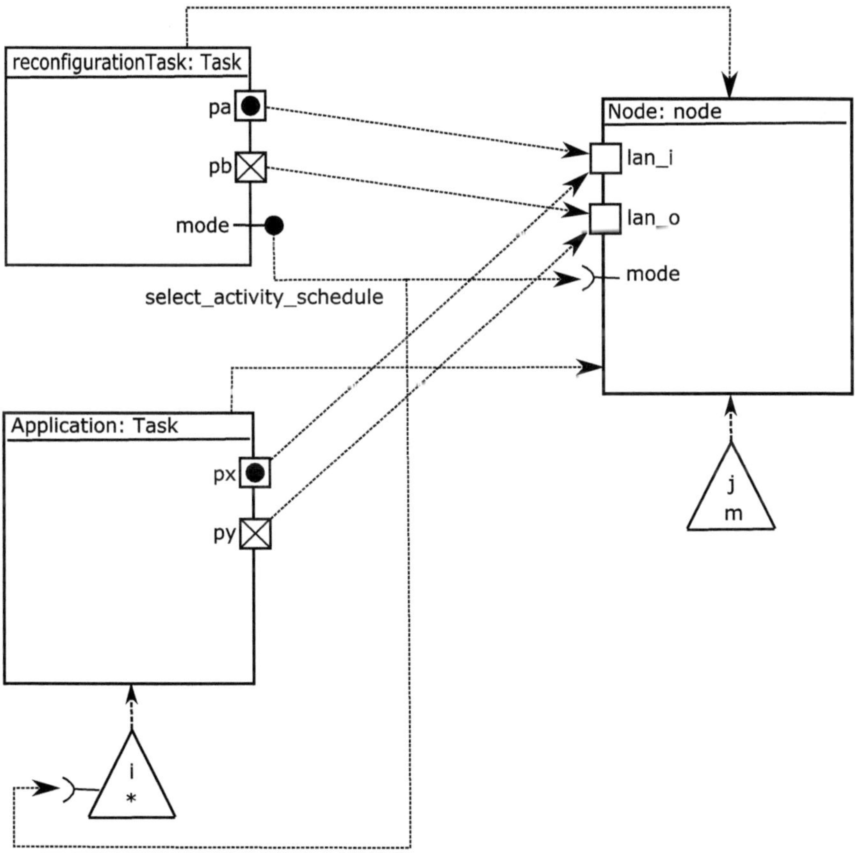

Fig. 2.3 Design time optimization based on a decomposition

2.3 Design Patterns for Reconfigurable Real-Time Monitoring and Control

The main purpose a system is to implement a certain functionality, which transform the input data stream (set) into an output data stream (set) according to the requirement specification (covering both functional and non-functional requirements). This functionality is called primary functionality (PF, for short). The reconfiguration—carried out as a response (counter-measure) to the influences listed above—makes changes in the PF and/or in the hardware configuration and/or in the assignment of the PF components to resources (Fig. 2.4). The separation of PF and the RF is the key for resource aware, time constrained runtime reconfiguration.

Following this separation-of-concerns approach, one of the most used reconfiguration "meta-model" is the MAPE-K, which is shown in Fig. 2.5. According to this approach the reconfiguration is carried out via a monitor–analyze–plan–execute cycle. The monitor activity derives the symptoms from the raw observations. The analysis step builds situational awareness and initiates the generation of the new system configuration. The planning step determines the sequence of actions, by which the current configuration is transformed into the new one. The last step is the timely execution of the plan. It should be emphasized that in this scheme the "Sensors" element in the figure collects information about the system target to reconfiguration (and typically they have nothing to do with the sensors applied in data acquisition for the main function of the system). The same applies to the "Effectors": the effectors are the means of making changes in the system configuration.

The reconfiguration process (i.e. the MAPE-K cycle) results in a sequence of reconfiguration actions. Such reconfiguration actions can be categorized as follows:

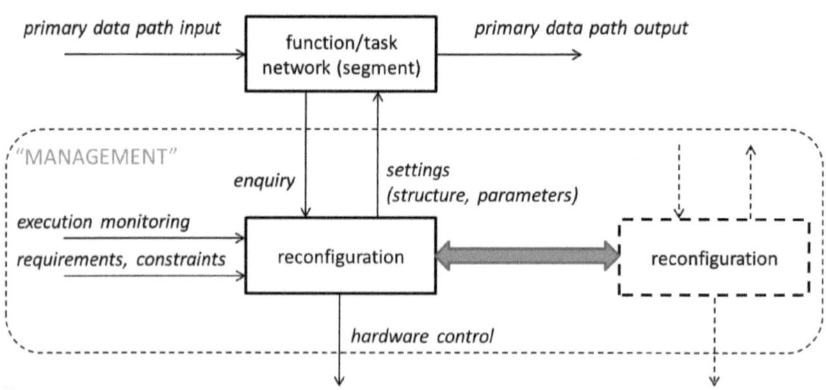

Fig. 2.4 The separation of PF and RF

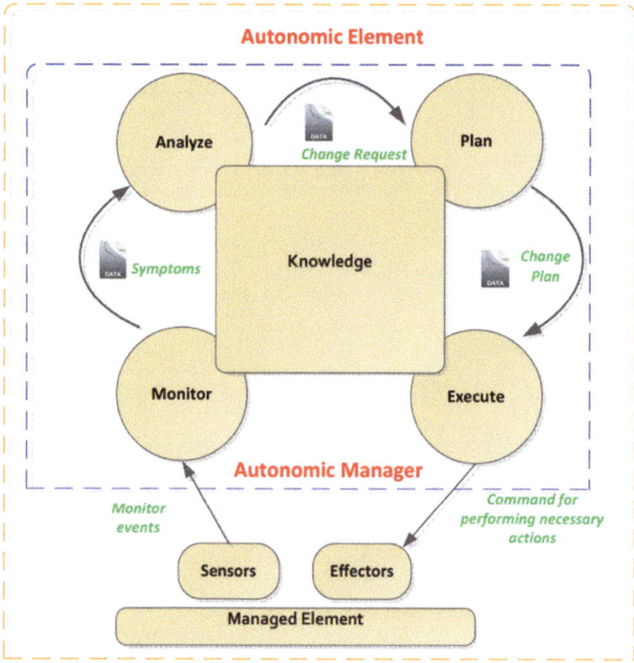

Fig. 2.5 The MAPE-K concept

Parameterization: During parameterization actions, the processing scheme remains the same except that only certain parameters of certain elements of the scheme are changed. For example, filter or protocol parameters. Conceptually this is the simplest form of reconfiguration.

Selecting algorithmic alternatives: Also here, the processing scheme remains the same. But certain elements are replaced by different realizations. The different realization has different non-functional properties, such us computation demand, information need, accuracy, robustness, etc.

Modifying the schedule of execution (of tasks): The processing scheme remains the same; indirectly the schedule influences the computing and communication demands, responsiveness, accuracy, priorities, etc., i.e. giving wide range of tuning capabilities.

Modifying the task allocation (task ⇒ processor assignment): Resource balancing, dynamic redundancy can be achieved via dynamic task assignment; typically the processing scheme remains the same, but re-parameterization is necessary in most of the cases.

Modifying the processing structure (task graph): This is the "deepest" reconfiguration activity; the result is a new processing scheme, which typically requires new task mapping and schedule.

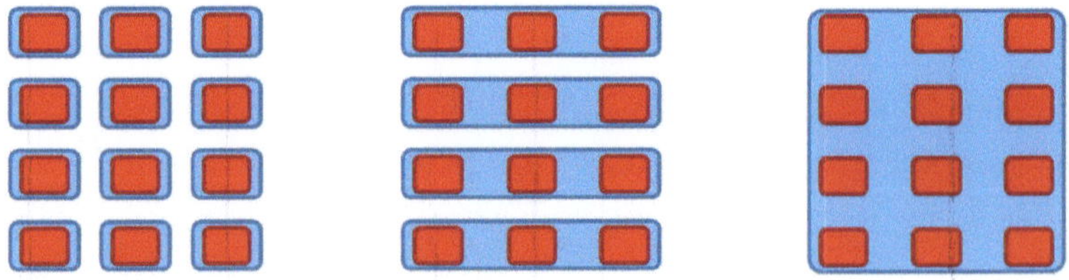

Fig. 2.6 The granularity of reconfiguration management

Reconfiguration can be carried out on different "granularity levels". At one extreme, the system binds a RF element to every PF elements. The RF element is responsible for "taking care" of the bound PF element, i.e. making sure that the operational conditions for the PF element is correct and the PF element properly contributes to the system level goals (Fig. 2.6a). This reconfiguration variant typically requires information exchange between RF elements, otherwise system level considerations cannot be handled.

The other extreme is having one single RF, which handles all PF elements. This is a centralized scheme: system level considerations can be handled, but information about the execution state of the PF elements—and their hosts in distributed cases)—should be collected and the actuation commands should be distributed—which may result in high communication demand (Fig. 2.6c).

An intermediate solution is when the RF elements manages a subset of PF elements (Fig. 2.6b). In distributed configurations a typical assignment is the node level reconfiguration management, in which each node hosts one RF element that controls all PF elements assigned to that node. Extra inter-node communication is necessary only to coordinate the actions among the RF elements.

In distributed implementations the "granularity alternatives" should be mapped into the physical configuration, and as such the interaction scheme between the RF is of primary concern. Besides that, the communication topology, the interaction scheme, and the allocation of RF components have direct impact on the performance of runtime reconfiguration. The factors listed also influence the performance of the primary data processing path as PF and RF functionalities compete for the same resources (processing/calculation capabilities, communication, energy, etc.).

The reconfiguration related activities include monitoring of the system state, building situational awareness, determining the new configuration, and carrying out the reconfiguration plan (actuation). At one extreme, all these activities rely on locally available information and actuation capabilities. Under a local scenario, the reconfiguration does not introduce communication inter-node overhead, and only local resource usage has to be considered. Typically, the extra processing needs of the reconfiguration related tasks. The local RF has a limited insight to the state of the whole distributed system, and thus the scope of the situational awareness and the reasoning about the optimal configuration is limited to the node itself. Only "selfish" decisions can be made about the reconfiguration, i.e. system level objectives typically cannot be achieved.

At the other end of the spectrum, the distributed management of the reconfiguration relies on shared situational awareness and cooperative "reasoning" about the updated configuration. In this case–at least theoretically–the optimal solution can be found[1] In summary, when designing a distributed runtime reconfiguration schemes, the following concerns should be explicitly addressed:

- The costs of building shared situational awareness: complexity of algorithms and communication overhead due to the distributed configuration.
- The costs of building coordination for concerted actions: communication overhead and the quality of the solution. For example, loosing the guarantee for finding the global optimum, relying on greedy or heuristic approaches.
- The costs of execution of RF: communication overhead, complexity of distributed planning, and actuation.

2.3.1 Formalizing the Reconfiguration Functionality

The runtime reconfiguration process can be considered as a continuation of the system design process that is performed at runtime. As such it remains a domain knowledge intensive process. Design decisions—such as algorithm selection, parameterization, interconnection topology, task allocation—should be made on various levels based on the actual circumstances, and guided by design expertise, general systems engineering knowledge, etc. Also when dealing with this complexity aspect, the architectural separation of the primary and the reconfiguration functionalities (with clear "actuation interfaces" between them, see Fig. 2.4) results in designs with cleanly assigned responsibilities and manageable non-functional properties.

The implementation of the reconfiguration functionality can take various forms. A frequent implementation pattern is to develop a custom algorithm addressing the particular case in hand. In this case, the "design knowledge" is hard-coded into the implementation. Such approach is frequently chosen for its runtime performance, but–due to the eventual complexity of the reconfiguration challenge–the solution can be error prone. In addition, extending this built in "design knowledge" can be an overly demanding work because it may require thorough rewrite and extension of the existing custom code. An more efficient and economic alternative is to follow

[1]Decomposing the design space exploration and the associated search/optimization into distributed configuration, that is a "divide and conquer" approach. Such sub-problems are significantly less complex than the whole, but interaction constraints should also be concerned and they are a challenge by themselves. See more details in the next section.

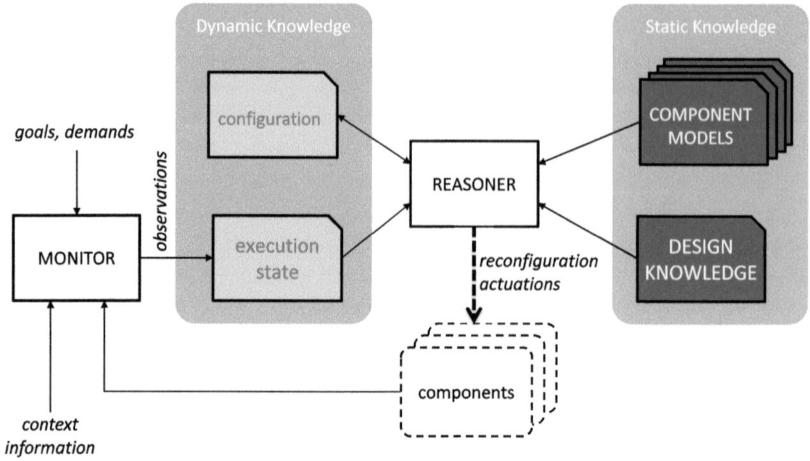

Fig. 2.7 A case independent reconfiguration scheme

the "knowledge based approach" pattern, in which the "knowledge" and the "use of the knowledge" are clearly separated entities.

Figure 2.7 shows a reconfiguration solution inspired by "knowledge based" pattern. As the name suggests the "knowledge based" pattern explicitly represents and uses knowledge. By knowledge, we mean any formal representation information, which is relevant in the context of making decisions about the using system resources to achieve the pre-set operational goals. In most of the approaches, the knowledge is captured into models that describe certain aspects of the operation of the system. Knowledge elements also describe the current state of the system relevant to reconfiguration: this part is called dynamic knowledge. Two main constituents of the dynamic knowledge are:

Configuration: This is the representation of the actual architecture the systems describing the components involved, their parameters, interconnections, etc. Differently stated, the configuration is a "machine friendly" representation of the design documents. It should be emphasized that

- the configuration does not necessarily describes the full design (which can be overwhelmingly complex) but only those parts and aspects, which are target of reconfiguration (see also the "runtime—deisgn-time tradeoff" considerations);
- the representation used should be "mutable", as the configuration description should be updated in order to reflect the changes introduced by the reconfiguration process

Execution state: The execution state reflects the conditions of the operation, which may influence the satisfaction of the functional and non-functional requirements set against the system. It is assumed that these conditions can change during the lifetime of the systems, consequently they can trigger reconfiguration actions.

The generic design knowledge is "coded" into the static knowledge part of the knowledge base. The static objective refers to the fact that this part does not change during the operation of the system.[2] Using a metaphor, the static knowledge is a set of textbooks a designer can use to acquire design knowledge.[3]

The main component of the reconfiguration scheme of Fig. 2.7 is the reasoner process, which "combines" these knowledge sources in order to satisfy the goals and demands under the given (current) circumstances. From the reasoner point of view the static knowledge is given, the execution state is derived from external sources—thus its scope of influence is making changes in the configuration. When the reasoner finds a new configuration, which satisfies all requirements, this configuration can be implemented, i.e. the changes are introduced in the physical configuration of the system by the reconfiguration actions. The reconfiguration loop is closed then via the monitoring of the system and updating the execution state information.

In this section, we use the "knowledge representation" and "reasoning" in a very wide sense: wide spectrum of representation and associated precessing mechanisms (reasoning) are available. There is no "ultimate solution". The selection of those should be carried out by thoroughly analyzing of the properties and needs of the problem in hand.

A few alternatives:

- Constraint satisfaction
- Optimization
- Search (explicit design space exploration)
- Feedback control (control theoretical formalization)
- Pattern matching
- Mathematical logic (theorem proving)

2.3.2 *Task Models for Runtime Reconfiguration*

Task models is a design pattern for reconfiguration extensively used in the DEMANES project. The idea behind task models is that the behavior of a (distributed) embedded system should be formalized as a set of interacting tasks. Consequently, the processing scheme implementing the runtime reconfiguration should be mapped into a task model. Each reconfiguration scheme define roles for tasks and fixes where they

[2]In this chapter, we do not consider learning systems. Learning systems are a subset of adaptive systems, which are capable of improving their knowledge during operation based on experience or user feedback. For example they can introduce new design rules based on frequently occurring cases or modify design rules if their application creates unwanted behavior. In the design scheme introduced the learning would be manifested by changes in the "static knowledge" part, i.e. strictly speaking it could not be considered static anymore. Still, the mechanism, which would introduce changes is outside of the reconfiguration mechanism considered here, i.e. from this viewpoint the knowledge is read-only and conceptually static.

[3]Due to the non-learning scenario this knowledge remains constant, i.e. the extra knowledge a human designer may acquire via experience is not stored.

should be located. Here we consider only the most commonly used configurations, namely:

- Local monitoring, local reasoning, local actuation (LLL type reconfiguration)
- Full scale monitoring, local reasoning, full scale actuation (FLF type reconfiguration)
- Constrained monitoring, local reasoning, local actuation (CLL type reconfiguration)
- Constrained monitoring, constrained reasoning, local actuation (CCL type reconfiguration)

The interpretation of the terms is as follows:

- Local: the scope of the activity is restricted to the node hosting the PF targeted by the reconfiguration (the PF mentioned here may only be a part of the complete PF)
- Constrained: the scope of the activity is restricted to the a subset of nodes hosting the PF
- Full: the scope of the activity covers the total system (i.e. all nodes are involved)

The LLL type of reconfiguration is the simplest scheme: every node monitors its own execution state, reasons locally about local goals and the reconfiguration actions are restricted to the node itself. The FLF type of reconfiguration corresponds the centralized implementation of the reconfiguration: one assigned node collects all execution state information from all nodes comprising the system, carries out the reasoning (locally) and actuates components on all nodes. The CLL differs from LLL that the local reasoning uses information about the execution state of a subset of nodes, typically neighboring nodes. The CCL scheme relies on cooperative reasoning mechanisms: beside sharing execution states the reasoners cooperate during the reasoning process to establish consensus, i.e. they attempt to achieve system-wide optimality instead of selfish local optimality.

The task models are very similar visually for all reconfiguration schemes, be it LLL, FLF, CLL, or CCL. The difference occurs in three points: the cardinality of the components, the connectivity between the reasoner tasks, and the mapping of the tasks to physical components. Expanding the task model and applying the mapping yields very different operation schemes.

Reconfiguration schemes using non-cooperative reasoning are common when using a local reasoning strategy (e.g. LLL, FLF, and CLL). Their task graph is shown in Fig. 2.8. In these cases, it is usual to have only one reasoner (indicated as reconfigReasoner) even when monitors and actuators are distributed in different nodes. The use of many reasoner tasks sharing monitors and actuators may happen. However, different reasoners do not exchange information and thus they do not cooperate. This can be noted in the task model diagram by the absence of communication links between reasoner tasks.

Reconfiguration schemes using cooperative reasoning are mostly common when using a distributed reasoning strategy (e.g. CCL). Their task graph is shown in Fig. 2.9. In these cases, there exist always more than one reasoner in the system

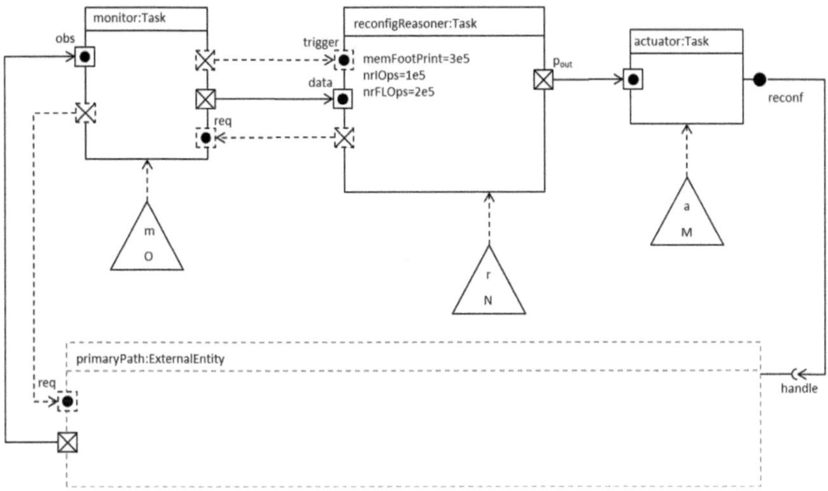

Fig. 2.8 Non-cooperative reasoning for reconfiguration

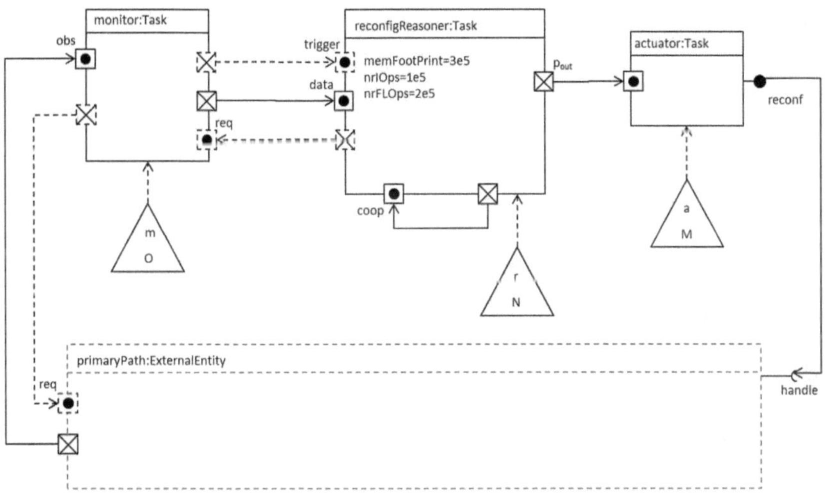

Fig. 2.9 Cooperative reasoning for reconfiguration

and they do communicate with each other. They may or may not share monitors and actuators, but decisions are taken with consideration of the direct information exchanged between them. In the task model diagram, this direct communication is denoted by the presence of communication links (self-loop in the reconfigReasoner block) between reasoner tasks.

Cooperative and non-cooperative reasoning can be used in all reconfiguration schemes discussed before. The differentiation between reconfiguration schemes is determined by the way the task to node mappings happen.

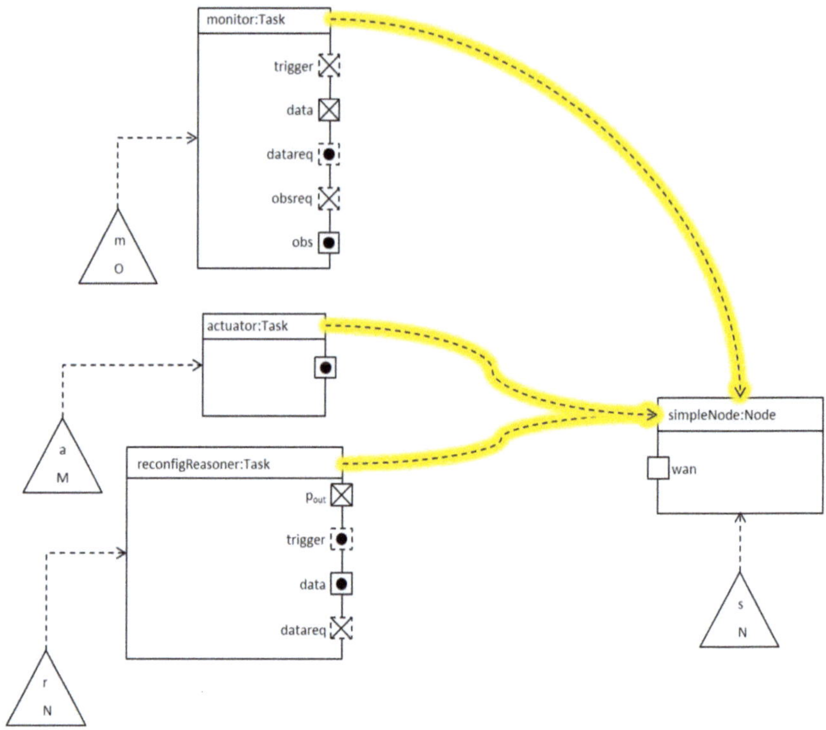

Fig. 2.10 Task mapping for the LLL reconfiguration scheme

2.3.2.1 The LLL Reconfiguration Scheme

When the LLL reconfiguration scheme is applied, reasoners that are mapped to one given node use only monitors and actuators that are mapped to this same node. That implies reasoning is made based on local information, and the scope of the reconfiguration action is restricted to the node itself. Most usually, only one reasoner is used in this scheme, but this restriction is not mandatory. When many reasoners are used in a cooperative way, one extra condition must hold to characterize an LLL scheme: reasoners mapped to one given node can communicate only with other reasoners mapped to this same node. That maintains the reasoning information local within the node scope. The task-node mapping is shown in Fig. 2.10. The filtering notation in the map guarantees the necessary conditions to create an LLL scheme.

2.3.2.2 The FLF (Centralized) Reconfiguration Scheme

The FLF reconfiguration scheme implies that information is collected by monitors spread all over the system and transferred to one single point, where the reasoning

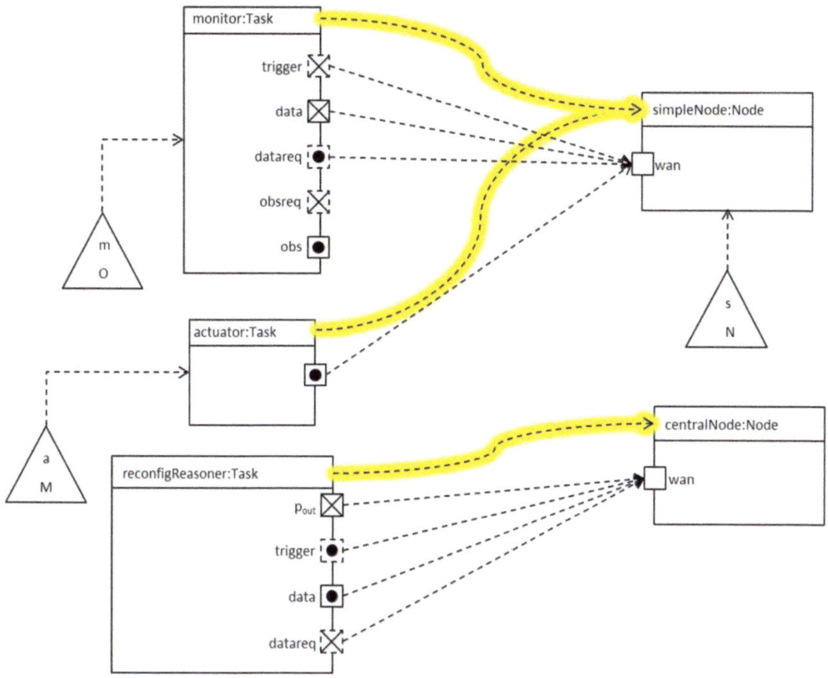

Fig. 2.11 Task mapping for the FLF reconfiguration scheme

takes place. Then, decisions taken by the reasoner(s) may trigger actuators placed in any part of the system. Therefore this scheme is usually interpreted as a centralized reasoning with global view and global actuation. When this scheme is applied, all the reasoners (normally one) are mapped to one single node, whereas monitors and actuators can be mapped to any other node. That implies reasoning is made based on global information (collected wherever necessary), and the scope of the reconfiguration action is also global (any node of the system). Most usually, only one reasoner is used in this scheme, but this restriction is not mandatory. That maintains the reasoning information local within the node scope. The task-node mapping for centralized schemes is shown in Fig. 2.11. The filtering notation in the map guarantees the necessary conditions to create an FLF scheme.

2.3.2.3 The CLL Reconfiguration Scheme

When the CLL reconfiguration scheme is applied, a reasoner that is mapped to one given node must only use actuators that are mapped to the same node—that implies their actuation scope remains local. However, such reasoner will use (at least one) monitors that are not mapped to the same node—that is, they will collect non-local information. This is typically the case when the reasoner has visibility to non-local

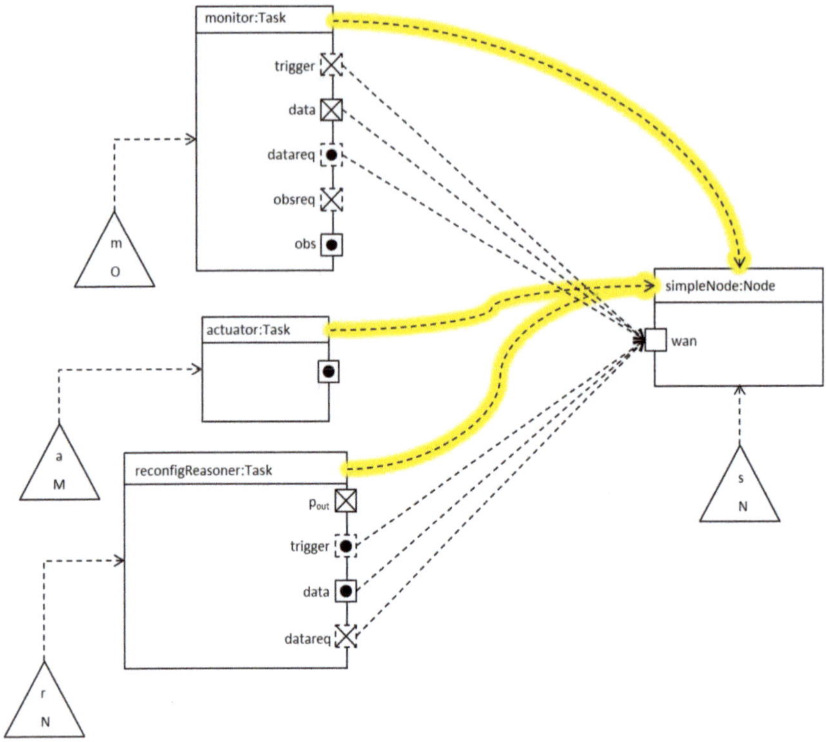

Fig. 2.12 Task mapping for the CCL reconfiguration scheme

information, but can actuate locally only. In other words, reasoning is made based on partial system information, but the reconfiguration action is restricted to a local scope. Most usually, only one reasoner is used in this scheme, but this restriction is not mandatory. When many reasoners are used in a cooperative way, one extra condition must hold to characterize the CLL scheme: reasoners mapped to one given node can communicate only with other reasoners mapped to this same node. That maintains the reasoning information local within the node scope. The task-node mapping is shown in Fig. 2.12. The filtering notation in the map guarantees the necessary conditions to create a CLL scheme.

2.3.2.4 The CCL Reconfiguration Scheme

When the CCL reconfiguration scheme is applied, a reasoner that is mapped to one given node must only use actuators that are mapped to this same node—that implies their actuation scope remains local. However, such reasoner will use (at least one) monitors that are not mapped to the same node—that is, they will collect non-local information. Also, they will exchange information with other (at least one) reasoners.

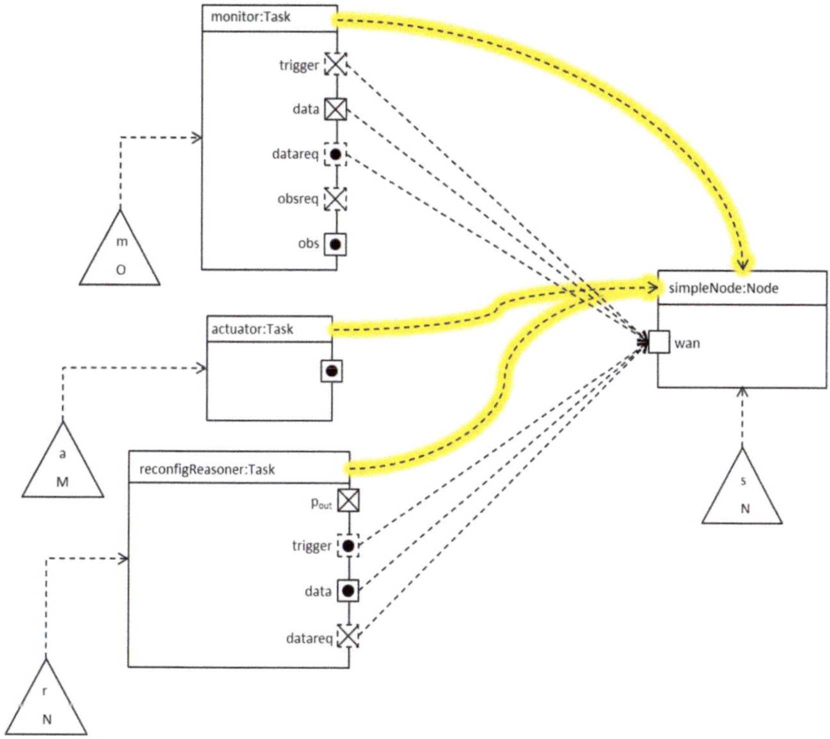

Fig. 2.13 Task mapping for the CCL reconfiguration scheme

This is typically the case when the reasoner has visibility to non-local information, cooperate with other non-local reasoners when making decisions, but can actuate locally only. In other words, reasoning is made based on partial system information and partial reasoning information, but the reconfiguration action is restricted to a local scope. For the CCL scheme, there are always more than one reasoner task. The task-node mapping is shown in Fig. 2.13. The filtering notation in the map guarantees the necessary conditions to create a CLL scheme.

2.4 Design Space Exploration for Runtime Reconfiguration

Design space exploration (DSE) is the process of searching through different system design alternatives. The aim of this process is usually to find some design which outperforms the other alternatives. Ideally one would want to find the optimal system design, but this is not always feasible. The power to operate on the space of potential design candidates renders DSE useful for many engineering tasks, including rapid prototyping, optimization, and system integration. The main challenge in DSE arises from the sheer size of the design space that must be explored.

The most challenging problems in design space exploration are managing the solution space size, and using a cost function which accurately describes what the desired performance is of the system.

2.4.1 A Quick Survey on Design Space Exploration and Design Decision Making

There are several techniques for performing design space exploration, be it at design time or at runtime. In the following, we summarize some of these techniques with example cases where they are applied for the design of embedded systems.

2.4.1.1 Genetic Algorithms

Palesi and Givargis [1] propose a method for exploring design space for a system-on-a-chip (SoC) architecture using a Genetic Algorithm (GA). They assume that the architecture is dependent on a set of parameters which have a large effect on the performance of the application run on the chip, but is difficult to tune manually. In order to run a GA it is necessary to define three things: A representation of the configuration (gene), an objective function or goodness measure and thirdly a convergence criterion which defines when a solution is found. In this example, and for any reconfiguration task, the represented gene is a vector of the parameters of the system. The function is in this case the power consumption and the time required to perform the task to be implemented. Finally the convergence criterion is a maximum number of iterations or a Pareto-optimal solution.

What a genetic algorithm then will do is iteratively try multiple variations of the configuration, and determine what the utilities (or fitness) of the proposed configurations are [2]. A selection is made from the top scoring configurations, and alterations are made upon these configurations for the next iteration. This method has some very clear links with the Darwinist theory of evolution. The exact method of altering existing configurations may vary from mutations where any parameter changes by a random offset, or crossover where complete sections are copied from other viable solutions. Some variants even allow for invaders which are completely random new competitors added to the pool at each iteration to guarantee a larger coverage in the search space area.

Using genetic algorithms has some advantages, namely that they are generic; little to no knowledge about the problem is required except for the three abovementioned things. Another advantage is that GA algorithms satisfy the anytime algorithm criterion: it will always come up with a solution, but the refinement improves as the algorithm runs for a longer period of time.

Downsides of the GA approach are the fact that it is not using any intelligence to optimize the existing solutions. It in fact quite haphazardly generates and evaluates random variations of existing solutions. Due to this principle it may take a long while before it converges to an optimal solution. It can also be quite computationally expensive to generate and evaluate thousands of possible configurations.

A GA could also be used to just optimize an existing solution provided from a different strategy. If it is true that parameters of the system are not critically dependent, and it is viable to do some runtime exploration of the configuration space, it is possible to run a GA continuously. It could change the parameters of the existing system, and monitor the performance. Noted that this approach would work very slowly, it could provide very nicely adaptive systems.

A similar approach is employed by Liu [3] in which a framework is proposed for evolving distributed algorithms. An algorithm is proposed in which agents at each node can choose between a set of basic transition functions, and the employed method is chosen initially randomly, after which the different agents will self-evaluate and change their function if it no longer meets the requirements. It can change via random mutation or by selecting and implementing the function of the best performing neighbor.

2.4.1.2 Graph Based Methods

Georgas [4] already proposed a method for describing reconfiguration as graph based models. This method called ARCM (Architecture Runtime Configuration Management) is specifically aimed to improve the visibility and understandability of runtime adaptive processes while allowing for human input in the adaptation-control loop.

At the core of their method (or linchpin as the authors put it) is the architectural configuration graph. Each node in this graph is a configuration, and the edges are transitions with a specific set of conditions. Whenever something is changed in the configuration of the system, this is stored in a graphical representation, and the amount of times the configuration is chosen, as well as the amount of time spent in that configuration is stored. Afterwards this graph can be inspected by the user in order to see how the system (re)acted.

Finite state machines have been used to describe reconfiguration by Teich and Köster [5]. They introduce a concept of self-reconfigurable finite state machines where in iterative steps a finite state machine can reconfigure itself by changing at most one transitions/output pair. In order to change a complete chain, some intermediate temporary configurations are required. If each intermediate configuration is considered a city, then the problem of reconfiguration in this case becomes comparable to the traveling salesman problem, and no algorithm can find an optimal way to reconfigure in polynomial time. Therefore the authors use a Genetic Algorithm approach to find a way to solve the reconfiguration problem. It should be noted that in this case the actual application that is reconfigured is a finite state machine, whereas the reconfiguration problem is not a finite state machine.

An example of a reconfiguration problem solved using graph matching methods has been shown by Kuo and Fuchs [6]. In their article the authors use reconfiguration using graph matching for designing large electronic circuits. Specifically the faced challenge is to allocate spare units in a way to optimally replace faulty units. Especially when multiple spare units are candidates to replace a faulty unit, the amount of possible reconfigurations can be very large. In the proposed solution the authors can solve the problem for allocating a spare unit to every faulty node in polynomial time O(V3) using the Hungarian Method [7]. This algorithm is still the most efficient for finding an optimal solution for a weighted bipartite graph matching algorithm. For unweighted bipartite graph matching problems there exist more efficient algorithms, most notably the Hopcroft-Karp algorithm which has a worst case complexity of O(V2.5), but for random graphs it runs in near-linear time.

2.4.1.3 Constraint Based Methods

Kogekar et al. [8] presented an approach for constraint-guided software reconfiguration in sensor networks. They have implemented a system which relies on monitoring the system requirements expressed as formal constraints. Those constraints drive the reconfiguration process that takes place in a base station that can communicate to all the sensor nodes. They subsequently demonstrate their approach using simulation results from a simple one-dimensional tracking problem.

In a related article from Eames [9], a method called DesertFD (Design Space ExploRation Tool using Finite Domain constraints) is demonstrated for design space exploration based on constraint satisfaction problems. The authors specifically mention that the tool was initially produced for design-time configuration, but was later embedded in a runtime reconfiguration framework in order to do on-the-fly optimization. They combine design-time derived metadata from profiles, benchmarks and expert knowledge, with runtime information from monitoring instruments. The process is then periodically called, and evaluates and prunes the set of system configurations with the goal of selecting the proper configuration to deploy.

An example of a reconfiguration model based on constraint satisfaction programming is given by Syrjänen [10] who examined of software configuration management. A declarative rule-based formal language is proposed for representing configuration knowledge. A case study is shown for managing Debian GNU/Linux packages. The author states that finding a stable in any program in normal logic is NP-hard. In order to find a solution for the configuration problem a method is used to find stable models semantics for logic programs developed by Simons [11]. This method called *smodels* is an algorithm for solving a constrained design problem, and as the authors denote, it does have some overhead and even possibly more than comparable methods. However this overhead provides the designer with a more powerful language. Consequently, problems that can be more compactly represented by logic programs can be more quickly solved with *smodels* than with a satisfiability checker.

2.4.1.4 Logic Based Methods

A logic based reconfiguration method will rely on some type of logical reasoner to find a configuration that matches requirements in runtime. One example a first-order logic interpreter is Prolog [12]. Prolog uses backtracking methods and Selective Linear Definite clause resolution to find solutions [13, 14]. This way of solving logic statements is proven to be both sound and complete [15]. Backtracking is the process of finding a solution by searching for partial solutions. For each candidate of a partial solution the validity is checked, if at one point no solutions exist for the last partial solution, a different one is searched for the previous one. This is analogous for finding solutions in a search tree, and looking depth-first for a solution that satisfies all sub statements.

Selective Linear Definite clause resolution is a method for checking statements by denoting them as disjunctive literals, and refuting all negated literals via the before mentioned backtracking method. SLD-resolution is very efficient both in terms of time and space. However, similar sub goals may be derived multiple times if the query contains recursive calls. Moreover, SLD-resolution is not guaranteed to always terminate [16].

In the SOSE project [17] a framework is described for reconfiguration at the heart of which lies a Prolog based reasoner. Using a relatively simple set of rules the authors can describe the reconfiguration strategies. The method is demonstrated for a greenhouse use case in which an optimal configuration needs to be employed for estimating the temperature distribution. Whereas in this case the goal of the system was purely state estimation, the framework was set up in a generic manner, so different applications are possible as well.

Isermann [18] already used a fuzzy logic controller for automatic supervision and fault diagnosis. He concluded that fuzzy logic provides a systematic framework to process vague variables and/or vague knowledge. As an example of this type of framework, Aubrun et al. [19] used fuzzy logic for monitoring and reconfiguring a power plant. They concluded that fuzzy logic provides better results for failure detection in terms of the robustness of the models comparing to classical methods. A freely available fuzzy logics reasoner called FuzzyDL [20] uses algorithms that combine a tableaux algorithm and that of a MILP-solver. It extends normal description logic with concepts of real values, integers, strings and definitions of fuzzy membership functions. It also allows for concepts in the knowledge base to be constrained using syntax which is comparable to constraint description. In a related article, the same authors propose a method for finding solutions using fuzzy logics from models described in OWL [21].

2.4.1.5 Fuzzy Logic-Based Decision Making

Fuzzy Logic is a powerful tool or framework for decision making in situations of essential uncertainty in models, information, objectives, restrictions and control actions, by emulating the making decision process of humans.

In the specific field of control systems, it is often the case in processes that control strategy designed according to other patterns (classical, adaptive, etc.), do not provide the desired results or simply fail. This is highly complex processes and yet, in the vast majority of cases, skilled and experienced human operators achieve satisfactory results. In 1965, Zadeh gives birth to the so-called theory of fuzzy sets, which lays the foundation of the so-called linguistic synthesis, showing how can be used vague logical statements to derive inferences (also vague) from imprecise data [22].

Linguistic Variables

One of the most important concepts in Fuzzy Logic is the linguistic variable [23]. In the classical viewpoint we only deal with conventional variables, and it is necessary for us to understand how variables have to be transformed before they can be handled by fuzzy systems.

The linguistic variable is characterized by a quintuple $(\chi, T_e(\chi), D, G, M)$ in which χ represents the name of the variable, $T_e(\chi)$ is set of linguistic values (attributes, adjectives) of χ, D the universe of discourse, G the syntactic rule to generate χ names, and M: the semantic rule to associate each value with its meaning. Therefore the variable χ can be transformed (mapped) into a "linguistic variable χ" and vice-versa.

For example, if *battery charge state* is interpreted as a linguistic variable (see Fig. 2.14), then T_e is the set of linguistic attributes defined for the *battery charge*. The fuzzy partition in the universe of discourse will be performed according to those attributes (*low, medium, high, very high,* etc.). For example:

$$T_e(battery\ charge\ state) = \{low,\ medium,\ high,\ \ldots\}$$
$$D(battery\ charge\ state) = [500, 2500]$$

In this case the use of the syntactic rule is not necessary because the attributes follow a natural order. However, in other cases attribute location needs to be arranged, e.g.:

$$T_e(battery\ charge\ state) = \{fairly\ high,\ excessively\ high,\ very\ high,\ \ldots\}$$

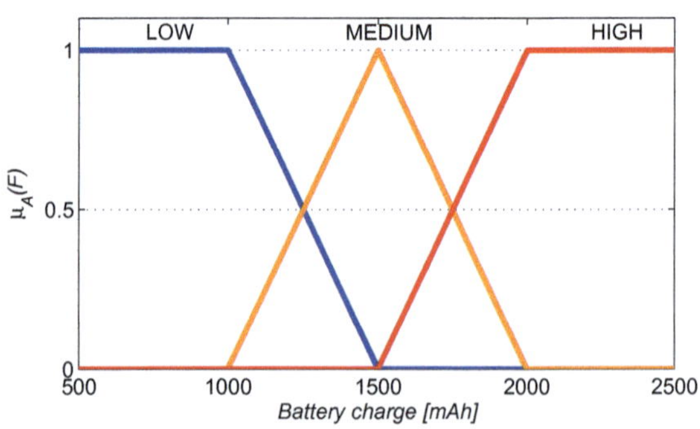

Fig. 2.14 Linguistic variable battery charge state

Then the syntactic rule could be: *G:..., "very high"* at the upper end of D, and after that *"excessively high"* and *"fairly high"*, respectively. The semantic rule M, for instance, allows us to interpret in our example that the battery charge is classified completely (100 %) as *"high"* above about 2000 mAh, the value at which membership in this set declines linearly down to zero at 1500 mAh.

$$M_{HIGH} = \begin{cases} \mu_{HIGH} = 1, & E \geq 2000\,\text{mAh} \\ \mu_{HIGH} = \frac{E}{500} - 3, & 1500\,\text{mAh} \leq E \leq 2000\,\text{mAh} \\ \mu_{HIGH} = 0, & E \leq 1500\,\text{mAh} \end{cases} \quad (2.1)$$

$$M_{MEDIUM} = \begin{cases} \mu_{MEDIUM} = \frac{E}{500} - 2, & 1000\,\text{mAh} \leq E \leq 1500\,\text{mAh} \\ \mu_{MEDIUM} = 4 - \frac{E}{500}, & 1500\,\text{mAh} \leq E \leq 2000\,\text{mAh} \\ \mu_{MEDIUM} = 0, & E \leq 1500\,\text{mAh};\ E \geq 2000\,\text{mAh} \end{cases} \quad (2.2)$$

$$M_{LOW} = \begin{cases} \mu_{LOW} = 1, & E \leq 1000\,\text{mAh} \\ \mu_{LOW} = 3 - \frac{E}{500}, & 1000\,\text{mAh} \leq E \leq 1500\,\text{mAh} \\ \mu_{LOW} = 0, & E \geq 1500\,\text{mAh} \end{cases} \quad (2.3)$$

Fuzzy Logic Device (FLD)

Zadehs conclusions suggested using a fuzzy rule-based (human reasoning-based) approach to the analysis of complex systems and provided a decision-making procedure together with a mathematical tool [24]. In general fuzzy systems are knowledge-based systems that can be built up from expert operator criteria and can be considered universal approximators where input/output mapping is deterministic, time invariant and nonlinear [25, 26].

The fuzzy logic device (FLD) is a general concept in which a deterministic output (crisp values) is the result of the mapping of deterministic inputs, starting from a set of rules relating linguistic variables to one another using fuzzy logic. For the mapping to be performed, deterministic values are converted into fuzzy values, and vice-versa. A FLD is made up of four functional blocks: fuzzification, the knowledge base, decision-making and defuzzification.

The classic configuration universally accepted as representing the four functional blocks described above is shown in Fig. 2.15 It is important to understand that these blocks have a functional meaning, so it is not necessary to separate them algorithmically or physically, as might be concluded from the diagram.

Regarding its input-output mapping, a Fuzzy Logic Device (FLD) has severely nonlinear input/output characteristics. The static characteristic of a typical two-input (x_1, x_2), one-output (y) FLD is shown in Fig. 2.16. The shape of the surface (in general

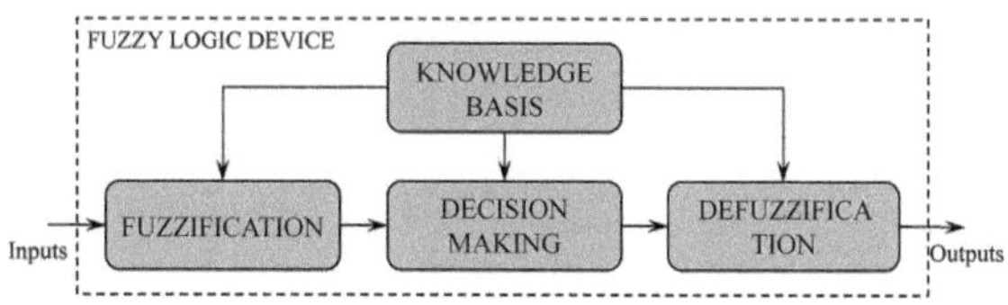

Fig. 2.15 Block diagram of a FLD

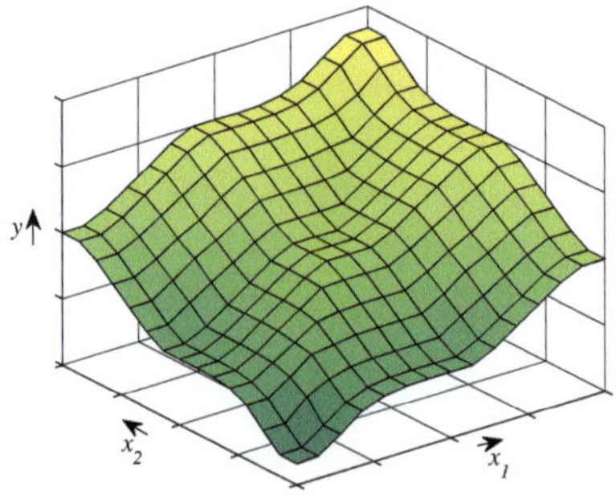

Fig. 2.16 Static characteristic of a two-input/one-output FLD

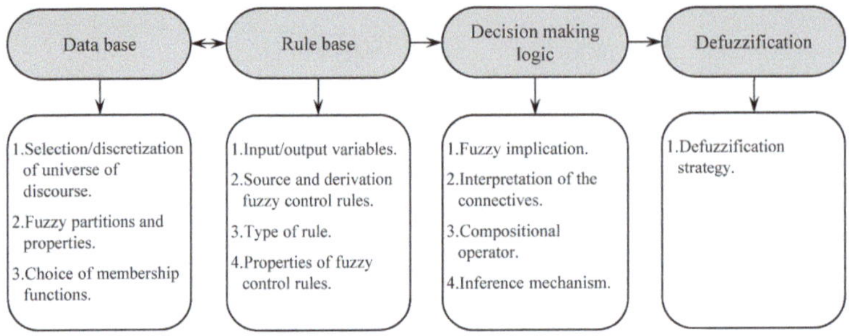

Fig. 2.17 Main FLS design parameters

a hypersurface) depends on the knowledge base. There are many parameters involved in designing an FLD, and they determine the shape of this nonlinear hypersurface.

The need to define linguistic variables by means of the FLD's input/output variables causes there to be a high number of parameters (design parameters) with usable information about the fuzzy partitions. The definition of operations in the decision-making procedure provides additional design data. So, there is enormous flexibility in the design stages, but the drawback is that these parameters have to be properly tuned. The main FLD design parameters are illustrated in Fig. 2.17.

One of the most significant steps is the selection of input/output, membership functions and the fuzzy partition of universes of discourse. The fuzzy partition of input/output spaces is what determines the "granularity" of the FLD. A heuristic procedure is usually applied to find the optimal fuzzy partition, although some algorithms based on neural networks and genetic algorithms are currently available for this purpose as well [27, 28].

However, the main step in FLD design is undoubtedly the derivation of fuzzy control rules. Tong [29] classified the ways for creating rules as *verbalization, fuzzification* and *identification*. *Vebalization* technique, assumes that the rule basis is formulated from verbal process descriptions obtained from skilled operators and technologists [30, 31]. *Fuzzification* (not to be confused with the fuzzification stage of the FLD) provides fuzzy decisional models drawn from ordinary (classical) mathematical expressions and using Zadehs Extension Principle. *Identification* involves the generation of relational descriptions obtained from numerical data gathered during manual process operation performed by expert operators. This method provides the objectivity of actual measurements, after processing large data files containing the information gathered during manual operation. Several computational procedures can be applied to implement *identification*; two good examples are *Fuzzy Clustering* and *Artificial Neural Networks* [32].

2.4.1.6 Rule-Based Ontology Reasoning

There are several mechanisms that can be used to define rules and manage them: to employ a semantic reasoner, a rules engine or a reasoning algorithm.

Semantic Reasoners They are also known as reasoning engines, because they use an inference engine to infer or deduce logical consequences from a set of facts or axioms. They are called semantic because they use a semantic language for reasoning or inference (OWL). OWL axioms infer new knowledge through language itself. Apart from infer new knowledge, the reasoners are also used to validate an ontology, i.e., see if it can create the instances that the developer needs to implement.

Rules Engine These systems are based on initial information and a set of rules, detects which of these rules should be applied at a given time and what results from its application. Describe the policies, standards, operations, definitions and constraints of a particular environment. These systems are based on rules to infer new knowledge. There are semantic rules engines that used semantic languages as OWL, and rules engines that use no semantic languages.

Reasoning algorithm The algorithms are able to predict or infer the behaviour of a user or system, based on taxonomy trees that store past experiences of the individual or system. No used for this purpose any semantic and power mechanism, it is based on the power of the implemented algorithm.

Semantic Reasoner

A semantic reasoner, reasoning engine, rules engine, or simply a reasoner, is a piece of software able to infer logical consequences from a set of asserted facts or axioms. The notion of a semantic reasoner generalizes that of an inference engine, by providing a richer set of mechanisms to work with. The inference rules are commonly specified by means of an ontology language, and often a description language. Many reasoners use first-order predicate logic to perform reasoning; inference commonly proceeds by forward chaining and backward chaining.

They are also known as reasoning engines, because they use an inference engine to infer or deduce logical consequences from a set of facts or axioms. They are called semantic because they use a semantic language for reasoning or inference (OWL). OWL axioms infer new knowledge through language itself. Apart from infer new knowledge, the reasoners are also used to validate ontology, i.e., see if it can create the instances that the developer needs to implement.

We can distinguish two types of reasoning:

Standard Inference It makes use of languages such as RDFS and OWL. Inference rules embedded in RDFS and OWL are fixed so that reasoners are able to implement inference even without the use of a rules engine, using only their own algorithms.
RDFS and dialects like OWL Lite and OWL DL are based on first-order logic. These languages allow through its expressiveness infer new assertions. To understand an example of reasoning that takes place, it possible to think of the class "Teacher", which is subclass of "Person". If the knowledge base contains an instance of Professor π, applying a reasoner against knowledge base and ontology, there would be a new assertion indicating that the instance of Professor π is a Teacher and a Person.

Rule-based Inference This type of reasoning requires an inference language for representing rules and a rules engine.
When the expressiveness provided by the ontology specification languages is not sufficient for semantic reasoner, then it can define rules using rule languages as SWRL. In these cases, not only reasoners employ an inference engine but also a set of semantic rules to generate knowledge and make inferences. The responsibilities of each of these elements are as follows:

- The inference engines interpret and evaluate the facts in the knowledge base to provide an answer.
- The semantic rules allow expressing the behaviour of an individual within a domain.

Reasoners allow defining semantic rules using different languages with varying degrees of expressiveness, and the inference engine is responsible of executing those rules in order to obtain new knowledge. It is essential that the rules are well defined in order to generate new knowledge successfully.

Ontology Reasoners

Reasoners can be classified into two groups: semantic reasoners and logic programming reasoners. Description Logic Reasoners perform standard inference through languages like RDFS and OWL. The description logics can represent a knowledge base that describes a particular domain. This domain is represented by classes (concepts), individuals and roles (properties).

Knowledge representation is made through two elements, TBox, which describes intentional knowledge as concepts and definitions of roles and Abox, which uses the ontology terms to declare assertions about individuals.

The descriptive logic reasoners (DL) must be able to provide the following inference services:

- Validation of the consistency of an ontology. The DL reasoner must be able to ensure that an ontology doesnt contains contradictory facts.
- Validation of compliance of the ontology concepts. The DL reasoner must determine whether a class may have instances. If a concept that does not satisfy this principle, the ontology is inconsistent.
- Ontology classification. The DL reasoner computed from the axioms declared in the ontology T-, relationships of subclass between all concepts explicitly declared to build the classes hierarchy.
- Details in the hierarchy concepts. The DL reasoner can infer what the classes which directly belongs.

Review of algorithms for runtime rescheduling/mapping

Dynamic migration of functionality from one device to another device is an important ability of self-adaptive reconfigurable networked embedded systems. There are two reasons why dynamic functionality migration is considered in networked embedded systems. The first one is reliability of these systems, e.g. when a device is added to the system or a device fails all functionalities are preserved. The second one is efficiency of the system expressed via an objective function(s) aimed at, e.g. energy consumption minimization or reliability maximization. Functionality in the systems is described via a set of tasks and functionality migration is realized by dynamic mapping and scheduling of these tasks with respect to actual state of the system. Mapping in this context means assignment of tasks to devices while scheduling means assignment of time slots when the tasks are executed.

The problem of task mapping is mainly addressed in parallel and grid computing area, e.g. [33]. Much less attention is paid to this problem in networked embedded systems. According to the quality indicator of task mapping expressed by the objective function(s) there are three possible approaches to task mapping and scheduling [34]:

Global approach There is only one decision maker (task mapper and scheduler) having single objective function.

Cooperative approach There are several decision makers (e.g. devices) that cooperate in making the decisions.

Non-cooperative approach There are several decision makers but each decision maker optimizes its own objective.

Majority of existing works deal with off-line mapping and scheduling of tasks considering global approach. Authors in [35] deal with task mapping and scheduling problem on networked embedded systems. Moreover, they also involved a control synthesis into the design process. The problem is solved off-line by a genetic algorithm. An ILP (Integer Linear Programming) problem formulation is proposed in [36]. This algorithm considers non-preemptive tasks and pipelining. Moreover, in order to exploit parallelism as much as possible replication of tasks is allowed.

Even less works are dealing with distributed algorithms for tasks scheduling and mapping. A self-organizing sensor network is described in [37]. The authors show a case study illustrating Kalman filtering on a distributed network of embedded systems. A distributed algorithm for on-line task mapping on reconfigurable networked embedded systems is described in [38]. The algorithm is based on diffusion algorithm, first introduced by Cybenko [39]. The disadvantage of the algorithm [38] is that it could generate huge communication traffic in the network. A diffusion algorithm is also used in [40]. Their algorithm is applied to a homogeneous system, i.e. system where capabilities of all devices are equal. Unlike the algorithm in [38] they consider integer granularity of tasks.

2.5 A Systems Engineering Process for Runtime Reconfigurable NESs

Developing large-scale real-time systems pose significant challenges and has been a target of intensive research both in academia and industry. These efforts resulted in a number of standards and recommended practice some of them relatively generic [41], others are dedicated to particular application areas [42]. The networked embedded systems domain (with the extra flexibilities, constraints and uncertainties brought in by the networked topology) further increases the complexity of the design. In order to cope with these uncertainties and improve the dependability of the networked embedded systems when deployed for large-scale, difficult to maintain and/or critical scenarios in a resource and cost effective way, the runtime reconfiguration promises a great advantages. Unfortunately the runtime reconfigurability brings in a new dimension for design complexity and well-established tools and methodologies prove inadequate to cope with these challenges [43, 44].

In the previous sections it was shown that model-based engineering (MDE) techniques for system design are becoming even more important for this class of applications–practically the thorough use of MDE techniques is the only reasonable hope to answer these challenges. MDE techniques assure that the designer can make informed decision during the design process and thus can minimize the effort spent on design iterations.

This section focuses on this decision process, i.e. it proposes a "work-flow" to get to a "proper" design in an efficient way. The main concern is addressing the added complexities brought in by the runtime reconfiguration capability. It is assumed that the requirements are well-defined and understood, thus the (far from obvious) requirement analysis and specification stages are not covered here.[4]

[4]Due to its complexity and far fetched consequences the requirement analysis and specification are thoroughly studied processes by themselves and long list of underlying processes, methodologies and tools (both commercial and research) have been proposed. Even a overview of the approaches would go beyond the scope of this book. Interested readers are referred to [45] for overview of frequently used approaches.

The programming/implementation stage of the system development only marginally influenced by the runtime reconfiguration: it is mainly the MDE approach, which makes a difference there. The impact of MDE approaches on system implementation is covered elsewhere (e.g. [46]). On the other hand the validation and verification of runtime reconfigurable demand for novel approaches, methods and tools. Because of its significance for the widespread introduction of runtime reconfiguration in practice, validation and verification are targeted in details in a separate chapters (see Part 2).

2.5.1 *Related Work*

In order to increase the efficiency of and reduce the uncertainties in the system development process development of methodologies and supporting tools is a target of intensive research. One of the first extensively published and studied system development family methods relied on the waterfall model. Tough the underlying process is clear and logical, these methods fail seriously in larger scale problems. It can handle the requirement and design uncertainties especially inefficiently: as a result the design feedback loops (i.e. making corrections in the design, choosing different alternative, etc.) are unacceptably long, effecting a number of development stages (Fig. 2.18). Still, the waterfall model is a "reference point" and later developments attempted to eliminate some of its shortcomings.

One of the other "mother of development methodologies" is the so called V-model. As far as the distinguished development stages are concerned the V-model uses the same concepts as the waterfall model, actually it can be considered as a "folded waterfall model". The significant difference is the emphasis on and the structuring of the different levels of validation and verification (Fig. 2.19) The V-model was definitely a step to the right direction, still the system level validation and verification can only be carried out late in the development process. Consequently–though

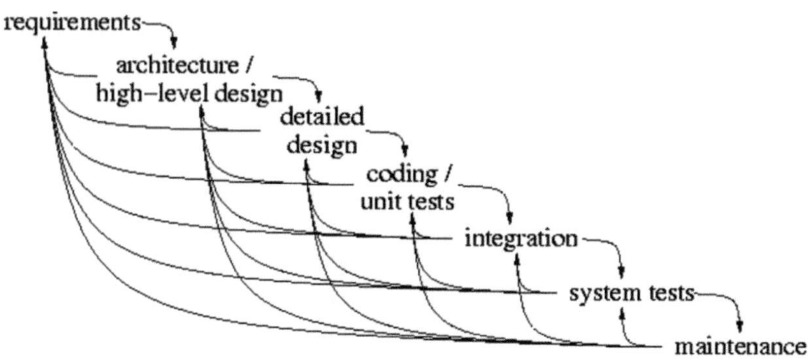

Fig. 2.18 Waterfal development process inreality

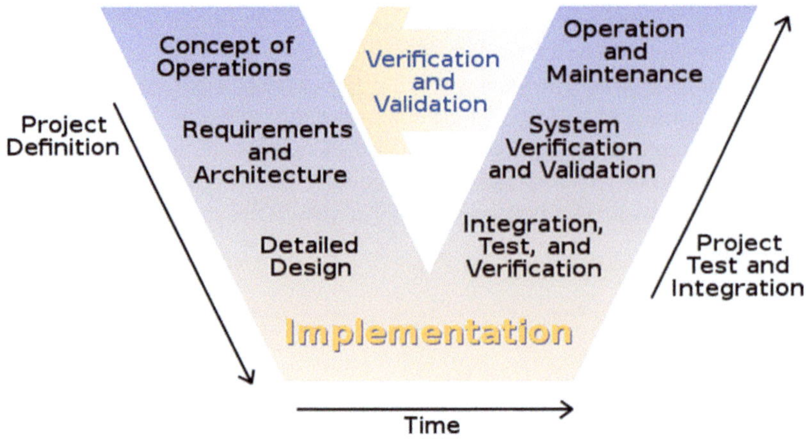

Fig. 2.19 The V-model of the system development process

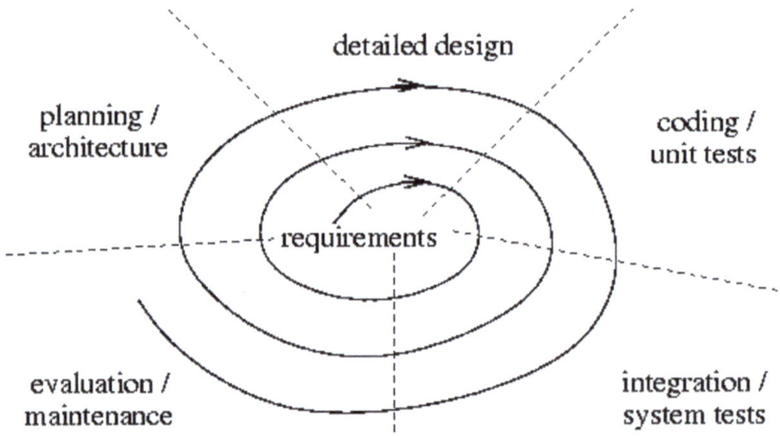

Fig. 2.20 System development in iterations

subsystem specifications considered correct, subsystem validation and verification completed–the complete system may not solve the real problem, i.e. there can be discrepancies with respect to the system level and end-user requirements. Similar issues of course may surface on lower levels of decompositions, too.

More recently developed methodologies attempts to shorter the feedback path, i.e. during the development more guidance is provided to the system designers and developers. Various iterative development process proposals were developed. A common development process structure is shared, as shown in Fig. 2.20. The underlying principle is that during the full development process a number of "complete solutions" are delivered and tested. The "completeness" is defined in such a way that the products can be shown to and experienced by the end-users, i.e. they can provide feedback about he product. The products delivered by the iterative development are typically extended variants of that of previous cycles, e.g. extended functionalities, improved user experience, tighter integration to legacy systems, etc.

The agile development methodologies brings the iterative approach to the extreme, requiring short and tightly scheduled iterations, particularly organized development teams and underlying business organizations [46].

Though iterative development processes (incl. agile methods) showed significant success in user centric and/or business software development in the (networked) embedded domain they have to be considered with caution. The tight integration of the developed system with the embedding physical processes, the new class of non-functional requirements (e.g. fault tolerance, temporal properties as condition of correctness, safety criticality, etc.) makes the specification, design implementation and testing/validation more demanding: the role of engineering disciplines is more pronounced, the validation and verifications are especially difficult/expensive/time consuming due to tight links to the embedding processes–rendering the rapid product deployment cycle infeasible. The model-based approaches come naturally as a rescue. The extensive and thorough use of models throughout the whole development process can provide frequent feedback, means of communicating with end-users, and (formal) verification of design decisions. A "pre-MDE" age development methodology emphasizing the role of modeling and design feedback is the SIMILAR systems engineering process [47]. The process is developed by comparing the similarities of several different systems engineering processes and extracting the similarities of the processes into SIMILAR process. SIMILAR is a dynamic, hierarchical, recursive and iterative process that evolves with time where several actions are done in parallel. SIMILAR process uses continuous re-evaluation functions and feedback loops as shown in Fig. 2.21. SIMILAR explicitly states the role of using models and the management of design alternatives. Also, the importance of continuous evaluation, validation and verification is emphasized by placing these activities to the center of the process and indicating the necessary interactions between the design stages. Though conceptually clean, the SIMILAR proposal cannot answer the challenge of speeding up the design process by shortening the design iteration loops and aiming for "first time right" designs.

2.5.2 The Customized Design Process

The runtime reconfiguration in networked embedded context makes the development process even more complex. The runtime reconfiguration renders the more traditional systems engineering processes inadequate and customization of the processes is needed to cope with the specialties of the runtime reconfiguration. The customized process

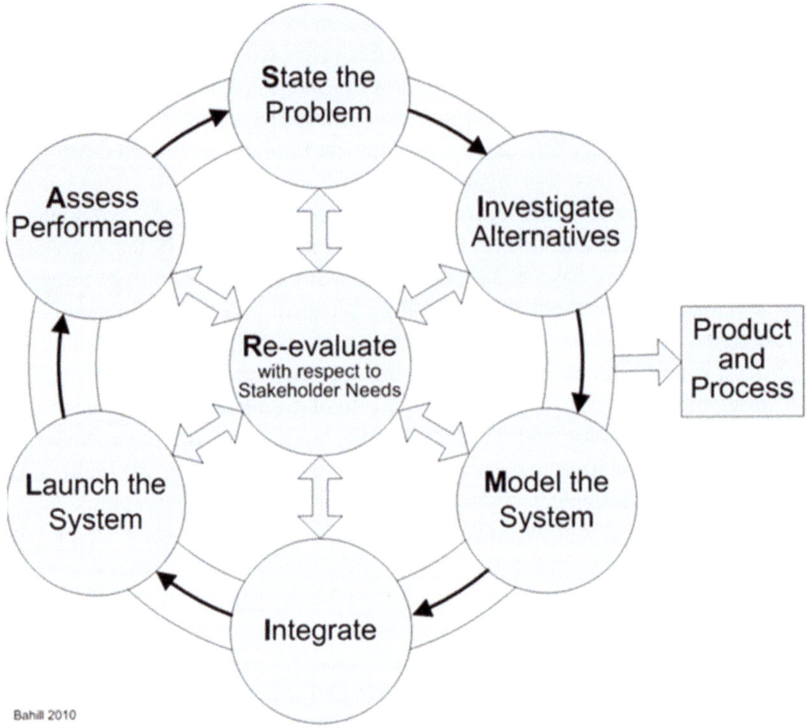

Fig. 2.21 The SIMILAR systems engineering process

- should implement an iterative design scheme to assure "continuous" feedback during the design;
- should facilitate the shortening of the feedback path, i.e. the design iterations should be quick: the "quick" adjective reflects both speeding up the feedback by quick completion of the individual design stages in the iteration and shortening the feedback loop (i.e. avoiding jumping back several stages in the design process) by facilitating informed design decisions;
- should support creating "architectural design patterns", i.e. enabling the use of proven design frameworks–especially for the integration of runtime reconfiguration mechanisms.

These requirements immediately indicates why the one of the "classic" processes won't work: the V-model falls short considering the "length" of the design feedback loop–which translates directly to time and costs. The iterative (incl. agile) processes are very difficult to realize due to the typically overly demanding non-functional requirements and tight integration with the embedding environment. The nominal incremental functional extension and testing cycle is not feasible to carry out in this context because the the design is typically not composable due to strong interactions among design aspects (see Sect. 1.3 for details).

In the customized design process the iterative scheme is implemented along a different dimension: instead of iterating by function extension, we iterate by adding details to the design. The main goal is to evaluate, test and verify the full system in every iteration. Every design iteration delivers the relevant emerging system characteristics (key performance indicators, KPIs), which can directly be compared to the system requirements. If the some of the KPIs are not compatible with the requirements, the design should be adjusted (design feedback loop). An iteration cycle's fundamental goal is to detail the design and derive more accurate KPIs. The challenge is to increase the detailing and accuracy of these steps in every iteration while minimizing the implementation work in order to assure speedy iterations.

Building the design process strictly around model-driven design approach is the most promising approach to achieve these goals. Ideally the design and implementation stages are tightly integrated: the actual implementation is the result of model refinements and automatic code generation. The frequent (almost "continuous") validation and verification is carried out in the model domain. The process can be represented as a modified iterative design process (Fig. 2.22). In each iteration the following steps are distinguished:

- Planning
- Design
- Modeling
- Integration
- Evaluation

Each iteration delivers a "complete system", ready for evaluation in the context of the system requirements. It can be said that essentially the designer traverses the left-hand-side of the V-model (i.e. moving from higher level design to implementation), but still the frequent check against specifications is assured (Fig. 2.23). Every design

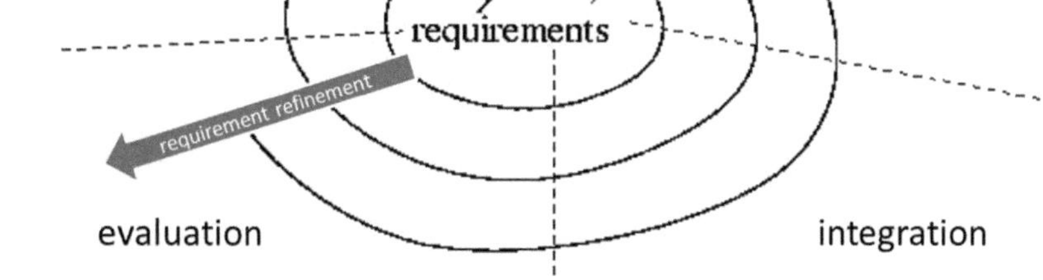

Fig. 2.22 Model driven iterative development process

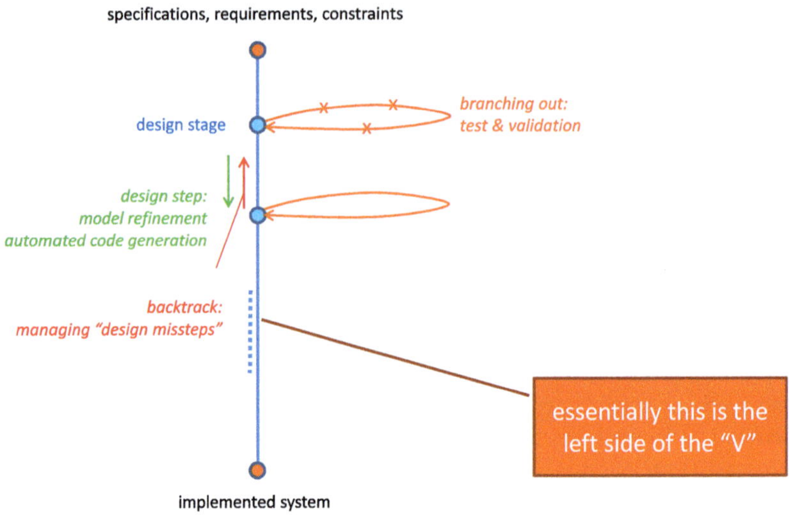

Fig. 2.23 From requirements to implementation

stage in the figure represents one full circle in the iterative scheme. As the design evolves, the next design stage is reached via completing design refinement of via automatic code generation.[5] In the model-driven context the refinement step covers[6]:

- Adding more details: This corresponds to the typical design decomposition activity. The designer "zooms in" on a (sub)system component and provides additional information about its composition. A typical examples are drawing a dataflow graph about a transformation element, which appears as a single "box" in the previous design stage of the system or providing additional attributes to a model element in order to facilitate higher fidelity evaluation.
- Adding new aspect(s): This refinement extends the scope of the modeling. Typically new aspect is introduced if a new class of requirements is to be managed. For example when the climate tolerance related requirements should be verified it will be necessary to include modeling aspects to cover heat generation (power) and heat transfer, heat radiation behavior.

At each design stage test and validation activities are carried out - see the integration and evaluation steps of the design iteration. It is important to emphasize that in parallel with the design refinements the requirements should also be detailed in order to allow more comprehensive testing.[7]

[5]In this section we focus on design refinement, i.e. when the designer further details the model to capture design and implementation decisions. Automatic code generation related concerns are covered elsewhere (see Sect. 3.2).

[6]For more details about model building methodology readers are referred to section ref-sec:1.6:methodology.

[7]In practice this is a very natural process: as the design is progresses so does the requirement specification detailing. The model driven development facilitates this aspect, too: when completing

2.5.3 Managing Runtime Reconfiguration

The needs for runtime reconfiguration emerge naturally when dealing with large-scale systems deployed in evolving environments: runtime reconfiguration can assure robust operation while keeping the system "lean and mean" and the design cost conscious. On the other hand implementing runtime reconfiguration capabilities makes the system design even more challenging.

As it was already overviewed in Sect. 1.4 runtime reconfiguration can be achieved via making two "extra functionalities" part of the deployed system:

- Building situational awareness, i.e. deriving the understanding about the execution conditions in the embedding environment and in the system itself (e.g. degradation of sensors, availability of resources, etc.);
- Managing of the situation, i.e. deriving action plan about achieving the goals as specified by the end-user and under the given situation.

The execution of these functionalities assumes the availability of sensing and actuation capabilities. The situational awareness relies monitoring of system states, which may not play role in the processing stages of the primary data path, i.e. dedicated sensing infrastructure may be needed.[8] The actuators (similarly to sensing) can be physical or conceptual) define the means for making changes in the configuration of the system: this can be parametric (e.g. changing parameters of algorithms (such as filter parameters)) or structural (e.g. changing communication topology, reallocating tasks, modify scheduling, etc.).

Finding the proper actuation (i.e. finding the new configuration) is the most challenging problem of implementing runtime reconfiguration. Note that the set of actuation points (and their value domain) corresponds to the free variables of the design thus it defines a design space, which should be explored during the operation of the system. Differently stated, certain design decisions are postponed to runtime and we need an "intelligent entity" to complete the design, i.e. find a reasonable (if not optimal) configuration for the given situation. Consequently in runtime

- the design space should be represented,
- algorithms should be executed to explore the design space,
- algorithms should be executed to assess the quality of the various design alternatives (generated during the design space exploration), and

(Footnote 7 continued)
a design iterations the results of the validation and verification can be communicated with the end-user, who can react with providing more detailed requirements or adjusting the existing ones. This is why working on the "complete system level" is important: this is the safest way to involving the end-users in the design cycles.

[8]This sensing infrastructure may (and typically does) consists of dedicated physical sensors (e.g. temperature sensor for ambient temperature, current measuring probe for power consumption estimation, etc.) and "virtual sensors", which are interfaces to internal system properties, which are available but are not considered on the primary data path (e.g. input signal level of the wireless receiver, CPU utilization, etc.).

- algorithms should be executed to guide the design space exploration, i.e. to find an optimal/reasonable solution.

Extreme caution is needed when selecting the actuation points because contradicting requirements:

- The number of actuation points and their value set defines the design space for the search for solution—which may easily become unacceptably large (exponential complexity). Consequently there is a need for minimizing the free parameters of the design.
- On the other hand the size of the design space determines the adaptation capabilities of the system: larger the space (i.e. more actuation possibilities) wider the operational envelop the system. This is definitely a incentive for introducing more actuation points.

These contradicting requirements should be balanced taking into account the availability of resources for executing the reconfiguration related activities and temporal constraints for reconfiguration. This is where the model driven engineering comes as a "rescue". Runtime reconfiguration related functionalities should be modeled as set of interacting tasks. Depending on

- the selected design space representation,
- the associated design space exploration algorithm,
- the dimensions and the size of the design space, and
- the selected search/optimization algorithm

the resource requirements of the reconfiguration can be estimated. Based on these estimations the reconfiguration related tasks in the task model can be properly parametrized and after determining the task mapping (Sect. 1.2) the system model is completed and ready for evaluation under end-user defined scenarios. It should be emphasized that besides taking the resource demand of the reconfiguration into account, the effectiveness of the reconfiguration should also be estimated and these "effectiveness figures" should be evaluated in the context of the system level requirements (e.g. required quantified system performance characteristics under user-defined set of failure modes).

2.6 Conclusions

One of the major challenges when designing software for complex systems is caused by a lack of a specific set of rules (methodologies). Adaptation (reconfiguration) to field conditions is difficult to model and implement on systems composed of a larger number of devices/components (distributed systems or systems of systems).

For state-of-the-art technology such as wireless sensor and actuator networks (cyber-physical systems), addressing the lack of a compressive set of rules for their design and realization offers considerable benefits. This leads to can accelerate and simplify their design and implementation.

With the extensions presented in this chapter, a model driven design approach becomes applicable to runtime reconfigurable cases. It should be noted that designing runtime reconfigurable system still remains a complex process. It starts with the extra difficulties requirement specification, continuous with the more demanding modeling and evaluation iterations and ends with challenging validation and verification. This latter has far reaching consequences for the whole design and will be considered in detail in Part 2. Model driven engineering fundamentally integrated into the development process is the key for the successful system design. We have to emphasize the importance of the careful balancing of the design-time–runtime trade-off and the importance of the high fidelity modeling of the situation assessment and management algorithms for implementing runtime reconfiguration. There are no "ultimate solution" and "design pattern" for runtime reconfiguration is general. Generic approaches can easily become unfeasibly complex thus taking into account the specialties of the case in hand and using proven heuristics are necessary. This emphasizes the importance of experimenting and thorough case-based evaluation. The proposed model driven development process can greatly speed up design iterations and reduce costs. Due to the design evaluation and testing challenges of networked embedded systems the seemingly extra effort needed for setting up and following strict model based design approach will be greatly compensated for during the later stages of the implementation, testing, deployment and maintenance. Modern systems engineering processes can easily be customized to accommodate for the specialties of runtime reconfigurable systems.

According to our experience the model centered methodology for building runtime reconfigurable systems is the main enabler for the widespread application of this technology.

References

1. M. Palesi, T. Givargis, in *Proceedings of the Tenth International Symposium on Hardware/Software Codesign, 2002. CODES 2002* (IEEE, 2002), pp. 67–72
2. D.E. Goldberg, (Addison-Wesley Professional, 1989)
3. Q. Liu, S. Dulman, M. Warnier, Area: an automatic runtime evolutionary adaptation mechanism for creating self-adaptation algorithms in wireless networks (2013). Under submission
4. J.C. Georgas, A. van der Hoek, R.N. Taylor, Computer **42**(10), 52 (2009)
5. J. Teich, M. Köster, in *Proceedings of the Conference on Design, Automation and Test in Europe* (IEEE Computer Society, 2002), p. 559
6. S.Y. Kuo, W.K. Fuchs, in *Proceedings of the 25th ACM/IEEE Design Automation Conference* (IEEE Computer Society Press, 1988), pp. 609–612
7. H.W. Kuhn, Nav. Res. Logist. Q. **2**(1–2), 83 (1955)
8. S. Kogekar, S. Neema, B. Eames, X. Koutsoukos, A. Ledeczi, M. Maroti, in *Proceedings of the 3rd international symposium on Information processing in sensor networks* (ACM, 2004), pp. 379–387
9. B. Eames, in *2006 IEEE Mountain Workshop on Adaptive and Learning Systems* (IEEE, 2006), pp. 127–132

10. T. Syrjänen, *A rule-based formal model for software configuration*, Technical Report (Helsinki University of Technology, 1999)
11. P. Simons, I. Niemelä, T. Soininen, Artif. Intell. **138**(1), 181 (2002)
12. OWL, http://www.w3.org/TR/owl-time/
13. K.R. Apt, R.N. Bol, J. Log. Program. **19**, 9 (1994)
14. M. Ben-Ari, *First-Order Logic: Logic Programming* (Springer, London, 2012), pp. 205–222
15. R.F. Stärk, *A Direct Proof for the Completeness of SLD-resolution* (Springer, Berlin, 1990), pp. 382–383
16. B. Gutmann, I. Thon, A. Kimmig, M. Bruynooghe, L. De Raedt, Theory Pract. Log. Program. **11**(4–5), 663 (2011)
17. C. van Leeuwen, Z. Papp, J. Sijs, in *16th International Conference on Information Fusion* (2013)
18. R. Isermann, IEEE Trans. Syst. Man Cybern. Part A Syst. Hum. **28**(2), 221 (1998)
19. C. Aubrun, D. Sauter, H. Noura, M. Robert, Int. J. Syst. Sci. **24**(10), 1945 (1993)
20. F. Bobillo, U. Straccia, in *IEEE International Conference on Fuzzy Systems, 2008. FUZZ-IEEE 2008* (IEEE World Congress on Computational Intelligence) (IEEE, 2008), pp. 923–930
21. F. Bobillo, M. Delgado, J. Gómez-Romero, Expert Syst. Appl. **39**(1), 258 (2012)
22. L.A. Zadeh, Inf. Control **8**(3), 338 (1965)
23. L. Zadeh, IEEE Comput. **21**(4), 83 (1988)
24. H.J. Zimmermann, *Fuzzy Sets theory and its Applications*, 3rd edn. (Kluwer, Boston, 1996)
25. W.A. Kwong, K.M. Passino, E.G. Laukonen, S. Yurkovch, Proc. IEEE **83**(3), 466 (1995)
26. P.U. Lima, G.N. Saridis, IEEE Trans. Syst. Man Cybern. Part B: Cybern. **29**(2), 151 (1999)
27. J.M. Bernardo, A.F. Smith, *Bayesian Theory*, vol. 405 (Wiley, 2009)
28. C. Jung, K. Kwon, Control Cybern. **27**, 545 (1998)
29. R. Tong, *The Construction and Evaluation of Fuzzy Models* (North-Holland, Amsterdam, 1979)
30. S. Kim, E. Kim, M. Park, Fuzzy Sets Syst. **81**(2), 205 (1996)
31. F. Matia, A. Jimenez, Int. J. Intell. Control Syst. **1**(3), 407 (1996)
32. T. Tobi, T. Hanafusa, Int. J. Approx. Reason. **5**(3), 331 (1991)
33. Y.K. Kwok, I. Ahmad, J. Parallel Distrib. Comput. **59**(3), 381 (1999)
34. D. Grosu, A.T. Chronopoulos, M.Y. Leung, in *Parallel and Distributed Processing Symposium., Proceedings International, IPDPS 2002, Abstracts and CD-ROM* (IEEE, 2002), pp. 52–61
35. A. Aminifar, S. Samii, P. Eles, Z. Peng, in *2011 IEEE 17th International Conference on Embedded and Real-Time Computing Systems and Applications (RTCSA)*, vol. 1 (IEEE, 2011), pp. 133–142
36. Y. Yi, W. Han, X. Zhao, A.T. Erdogan, T. Arslan, in *Design, Automation & Test in Europe Conference & Exhibition, 2009. DATE'09* (IEEE, 2009), pp. 33–38
37. J. Sijs, Z. Papp, in *2012 15th International Conference on Information Fusion (FUSION)* (IEEE, 2012), pp. 1012–1019
38. T. Streichert, D. Koch, C. Haubelt, J. Teich, EURASIP J. Embed. Syst. **2006**(1), 9 (2006)
39. G. Cybenko, J. Parallel Distrib. Comput. **7**(2), 279 (1989)
40. P. Neelakantan, Int. J. Comput. Appl. **39**(4), 7 (2012)
41. https://www.ibm.com/developerworks/rational/library/content/03July/1000/1251/1251_bestpractices_TP026B.pdf
42. ISO/IEC 107463 (1996)
43. G. Karsai, J. Sztipanovits, Intelligent Systems and their Applications, IEEE **14**(3), 46 (1999). doi:10.1109/5254.769884
44. T. Saxena, A. Dubey, D. Balasubramanian, G. Karsai, in *2010 Seventh IEEE International Conference and Workshops on Engineering of Autonomic and Autonomous Systems (EASe)* (2010), pp. 137–144. doi:10.1109/EASe.2010.22
45. B. Berenbach, D.J. Paulish, J. Kazmeier, A. Rudorfer, *Software and Systems Requirements Engineering: In Practice Education* (McGraw-Hill, 2009)
46. J. Whittle, J. Hutchinson, M. Rouncefield, Software, IEEE PP(99), 1 (2013). doi:10.1109/MS.2013.65
47. A. Bahill, B. Gissing, IEEE Trans. Syst. Man Cybern. Part C: Appl. Rev. **28**(4), 516 (1998). doi:10.1109/5326.725338

Chapter 3
Runtime Services and Tooling for Reconfiguration

Julio Oliveira de Filho, Teus Vogel and Jan de Gier

Abstract In model-based design methodologies, the system model is at the center of the development process. The model is the single source of design information, and typically is separated—logically and physically—in several views, such as requirements, specifications, design decisions, implementation guidelines, and testing protocols. When building up on such a model, the methodology prescribe—often formally—what the system model information represents, how it is manipulated and evaluated, and how it is refined from conception towards deployment. More often than not, system models and their manipulation tend to be too complex, formal, and multi-faceted to be made manually by a human designer. Such task requires appropriate tooling, that enables a correct-by-construction automation of the process. These tools then become a mandatory support to the designer. This chapter discusses how to build up a model-oriented tool chain and its most important activity flows: model editing, model visualization and transformation; model validation and evaluation; and finally, (semi-) automatic system implementation based on the system model. In addition and for each activity, emphasis goes also to the specific tool support which is necessary for the design of adaptive/reconfigurable systems.

3.1 Introduction: Model Oriented Tool Chain—An Overview

Figure 3.1 depicts the example of a design tool chain tailored to support model-based design. The central point of the design tool chain is the system model (represented as an XMI file) which is the single carrier of the design information. Around the

J.O. de Filho (✉) · T. Vogel · J. de Gier
TNO, Oude Waalsdorperweg 63, The Hague, The Netherlands
e-mail: julio.deoliveirafilho@tno.nl

T. Vogel
e-mail: teus.vogel@tno.nl

J. de Gier
e-mail: jan.degier@tno.nl

Z. Papp and G. Exarchakos (eds.), *Runtime Reconfiguration in Networked Embedded Systems*, Internet of Things, DOI 10.1007/978-981-10-0715-6_3

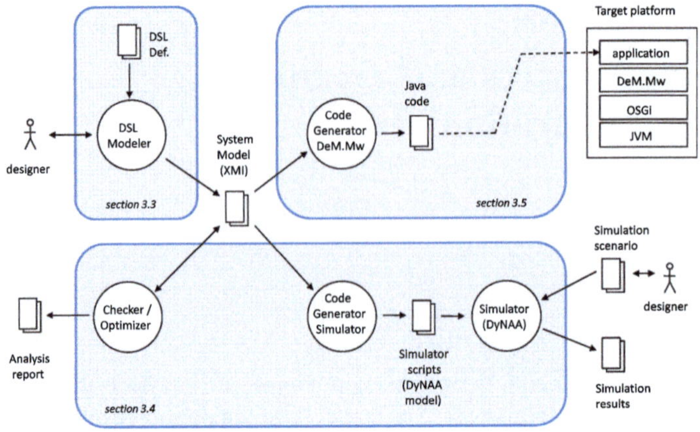

Fig. 3.1 Design tool environment

system model, design tools are grouped according to the activity they support during the design process.

The most important design activities and the tools frequently supporting them are:

System description and visualization Designers need tools for describing, editing, and visualizing their systems. And often such descriptions must express different views over the system, such as requirements, specification, physical organization, or functional decomposition, which may imply different tools. Tools in this group are therefore graphical and textual model editors, spread-sheet organizers, requirement assessment tools, or databases. This activity group is discussed in more details in Sect. 3.2.

Quantitative analysis, verification, and optimization Though each one of these activities is complex enough to have its own activity group, together they characterize the main work of the design phase: interpret and transform the model to generate design guidance. When performing these activities, designers require analysis tools, formal verifiers, simulators, validators, optimizers, syntactic and semantic checkers, etc. This activity group is discussed in more details in Sect. 3.3.

System transformation and implementation support When the design phase is over, an automatic translation to the implementation files/artefacts is most desired. That, to avoid that design details are lost or errors are introduced in case of a manual translation. Tools in this group are compilers, code generators, implementation architectural templates, and templates for design patterns. This activity group is discussed in more details in Sect. 3.4.

The fact that model-based designers assume a single model as carrier of the design information has two strong consequences to the development of design tools: First, tools are expected to ‘understand’ the model or part of it. As a result, automation processes and tools are frequently pushed to receive the model—as it is—as input;

Second, tools must often modify the existing model, aggregate information onto it, or create yet another system view.

The idea is to achieve—through this read, modify, enrich, and back annotate procedure—a more automatized and consistent design process. Modifications of the description by one tool, or newly added information propagate automatically via the model, that is, it is recognized or causes (visualization) update actions in other tools without manual interference. Consider for example a model visualization and editing tool coupled to a simulator. Results produced by the simulator maybe annotated back to the model—often during the simulation phase itself—and may be visible from the same graphical interface where the model is edited. The impact of new design decisions can be more quickly understood as modifications in the model automatically trigger visualization interfaces, verifiers, assessment routines, and test procedures.

Despite these advantages, a model-centric approach imposes serious challenges to the composition of a tool chain: First, tools have to keep pace with how the model is described, that is, with the evolution of the language/format used to describe the models. The way designers want to describe their models may change from time to time and dependent on the specificity of the system-under-design. Tools must be able to accommodate and deal with these variations. Second, tools have to be able to read, understand, modify, and write back the same model. Modifying tools to accept a certain input format is not always possible, specially when integrating third-party tools in the chain. Understanding the model data (e.g. its format) and information access management are typical problems found. These challenges increase when tools are developed independently or managed by different stakeholders.

The tool chain example suggested in Fig. 3.1 depicts also the authors' suggestions to cope with these challenges. Metamodeling technology(tools) is the key to deal with the introduction or evolution of modeling languages. Metamodelers ease and organize the formal specification of the modeling language—or metamodel. Also, they often offer facilities to automatically generate model editors and code generators that 'understand' the new languages. Ideally, modifications in the formal modeling language are transferred automatically or with less effort to editors and code generators.

Code generation and model transformation is the key technology to deal with exchange of data between models and tools. In Fig. 3.1, the editor outputs the model in a persistent XML format (an XMI file) which is formally specified by an XSD scheme. The model description, the metamodel, and the output (XSD) format specification can be used by code generators to extract model information and/or transform the model into the input format of other tools. Conversely, code generators can also be used to read the output of other tools and update the model description according to the format specification rules. In this way, system information can flow from the model to other design activities and back.

The following sections will discuss in more details the proposed tool chain.

3.2 Modeling Tools and Code Generation

The first group of activities in the design phase relates to describing the system model. Such description must be accomplished by using some sort of a formal language, which determines the syntax and semantics used to describe the model. The graphical modeling language introduced in Sect. 1.2 is an example of such a formal language which was intended to design dynamic reconfigurable networked embedded system. In model-based design approaches, the modeling language often supports the description of a system from multiple points of view—called aspects—such as requirements, specification, physical architecture, functional, etc. These aspects are interrelated, and together they form a consistent description of the target system.

When a modeling language is defined, a tool can be built to describe models by means of this language. Such tool is called a model editor. The model editor helps the designer to create machine readable models expressed in a certain modeling language, and to verify them against rules associated with the modeling language. In this way, a model editor supports the designer in describing a consistent and valid model. In many cases, such as in Sect. 1.2, modeling language concepts can be represented graphically, such that model editors can be implemented as graphical tools.

The output of the model editor is a persistent representation of the system model (e.g. a set of files), which can be stored by the tool itself or in a (central) repository, such as a file system or a database. Preferably, such persistent representation is also based on the modeling language. Stored models can be read, interpreted, modified, enriched, and transformed by other tools. For example, a model can be automatically translated into a set of human-readable system specifications (e.g. a generated report).

3.2.1 *Developing a Model-Based Modeling Tool*

In order to create tool support for a modeling language, one could choose to develop a graphical modeling tool from scratch, fully dedicated for that specific modeling language. This approach has a number of disadvantages: first, developing (and documenting) such a tool would involve a considerable amount of resources (money, man-hours) as well as specific programming expertise that might simply not be available. Second, once such a tool has been developed, modifying or extending the language becomes difficult and would involve a lot of effort. Finally, even when such tool is created, it still needs to be maintained, but guaranteeing access to the original developers for extended periods of time is not always an option. These drawbacks make the development of dedicated model editors non-viable.

An alternative approach is to use existing (graphical) modeling tool, that are to some extent reconfigurable (programmable) to support user-defined (graphical) languages. Several such tools exist, both commercial and in the public domain, and they come in different flavors:

- tools that need to be constructed by combining a set of available functions (e.g. in a building block manner). Support for the target language still involves a significant coding effort in a programming language, although a lot of the required functionality is made available through library functions. For example, in graphical frameworks /citeEclipse core graphical windowing functionality is already in place.
- tools that require the language designer to define the properties of the language in a purely declarative way. These tools are frequently called meta-modelers. Though meta-modelers are intrinsically tools to define and manage a modeling language (a meta-model), many of them are also able to automatically generate a model editor based on the language definition. Such model editor will have the language support a designer needs to describe the models.

In order to create a model editor for the system modeling language discussed in Sect. 1.2, the authors used the meta-modeling approach. In a general way and for this approach, a meta-modeling tool must meet the following set of requirements:

- The tool should support the development of modeling tools for graphical DSL's
- It should support the development of a domain specific language (DSL) containing multiple aspects
- It should support the construction of hierarchical models
- It should be possible to implement additional language constraints / rule checking
- It should support an import / export function to integrate it into a tool environment
- It should support user-defined text- and code generation from the system models

There are several options for meta-modelers in the market. In the following sections, a number of them are discussed shortly. With some of the tools, small modeling experiments were carried out. The experience obtained from conducting these experiments was taken into account when selecting the modeling tool to use in our tool environment.

3.2.1.1 GME

GME (Generic modeling Environment (GME) [1]) is a public domain tool, which belongs to the category of meta-modeling tools. GME is a configurable and extensible modeling environment developed at the Institute for Software Integrated Systems at Vanderbilt University. The most prominent feature of GME is its configurability. The GME toolset can be configured to describe, check, visualize, and process models in almost any graphical modeling paradigm. Such flexibility is achieved by using meta-models to guide the environment configuration. GME contains all functionality required for developing adequate support for the target language. However, the stand-alone version of the tool is no longer actively supported. Instead, development of a browser based client-server version is underway. This version was not available yet at the time of writing of this document.

3.2.1.2 Eclipse EMF/GMF

Eclipse [2] is a public domain IDE framework, for which a lot of 'plug-in' type modules are available. The Eclipse Modeling Framework, also known as EMF, is a project inside the Eclipse community that implements the meta-modeling support in the whole Eclipse platform. The EMF project is a very stable open source project, with a large community supporting it and a lot of projects built around it. Therefore, the EMF project has become the core of the Eclipse modeling module, and it is the cornerstone for many other projects, such as the UML2 project, the Graphical modeling Framework, etc. The EMF core module includes the main mechanisms to handle meta models, including an implementation of a reduced version of the Meta Object Facility (MOF), a meta-meta-model standardized by the OMG, called Ecore.

The GMF module is used to create the graphical support functionality. Creating all menus, pop-ups and rule-check involves a coding effort (in Java). Using Eclipse for creating and using a graphical modeling tool requires a long and steep learning curve. The GUI of the Eclipse IDE is complex and more tailored towards java coding. Although the EMF functionality works flawless, the GMF support was found to be buggy and ill-documented. Also, the visual quality of the created diagrams was rather poor. Creating language support for a user-defined language in Eclipse EMF/GMF was experienced as a laborious and painful process.

3.2.1.3 Papyrus

Papyrus [3] is a graphical modeling tool supporting UML-2, SysML and MARTE. Papyrus is integrates with the Eclipse framework and was implemented as an eclipse plug-in, using the EMF/GMF functionality. Using Papyrus, a language designer can specify its own domain specific language extensions, and defining its own DSL's based on the UML-2 and using the UML-2 profile mechanism.

3.2.1.4 MetaEdit+

MetaEdit+ [4–6] is a commercial meta-modeling tool, developed and marketed by MetaCase Inc., Finland. MetaEdit+ is actively supported and accompanied by good documentation. It incorporates all functionality necessary to create adequate support for a target modeling language, inclusive the definition of multi-aspect languages and the creation of hierarchical models. By using MetaEdit+, designers can create graphical diagrams of high visual quality. MetaEdit+ also supports rule checking and has a built-in interpreted language that can be used to implement language specific report generators and code generators.

It has been our experience that defining a graphical modeling language using MetaEdit+ is relatively easy. Writing code generators and performing advanced rule-checking involves more effort, but can be achieved using the built-in interpreted report generation language. Defining a DSL in MetaEdit+ is an incremental process with which high quality graphical modelers can be created with a relative small development effort. The models created using MetaEdit+ are stored in the tool's repository using a proprietary format. Code generators were implemented to produce an XMI description of the created model(s). Due to the available support, active development, high quality documentation and extensive functionality, MetaEdit+ was finally selected as the platform for creating the graphical modeling tool for the target language.

3.2.2 Meta Modeling

A modeling language is usually defined in terms of a set of object types and a set of relation types that can exist between the object types. Object types and relation types together are specified in a schema. Subsequently, the models that can be created according to this schema consist of objects and relations, where the objects are instances of the defined object types and the relations are instances of the defined relation types. The created models must conform the schema, that is, relations between objects can only be created if the associated relation type is defined between the object types with which the relation connects.

A model created in such a system can be viewed as a graph, where the objects are the vertices of the graph and the relations are the edges. A relation type has a set of roles specified. Some meta-modeling tools support only two roles per relation type (supports only binary relationships), while other meta-modelers support more than 2 roles per relation type (N-ary relationships). The roles in a relation type specify which type of object the relation can be linked to (for that) role. By using roles, it is possible to create richer models and to capture more real-world semantics of the domain that is being modeled.

Object types also can have a set of properties defined. The associated object will then carry values for each of the defined properties. Properties can be used to capture even more real-world information in the models. In some meta-modeling tools, roles are just special cases of properties, while in others, there is strict separation in the definition and handling of these two modeling elements. Figure 3.2 depicts a screen shot of a language definition session in MetaEdit+.

Because all models conform to the same explicitly defined schema, it now becomes possible to (mechanically) interpret and navigate the models and extract selected information. The extracted information can be used to verify that certain rules/constraints are satisfied, generate text (e.g. report and code generation) or create new models according to another schema (model transformation). The concepts and rules that are used to define a schema, are typically built-in in the modeling tool. These concepts are used to describe a user defined language at a higher (e.g. type-) level, hence the name 'meta'-modeler. In most meta-modeling tools, the built-in meta-schema is also self-conformant, that is, there exists a population of the meta-schema (a meta-model) that describes the meta-schema itself. This is not only just elegant, it can also be very useful because it allows all specification levels (including

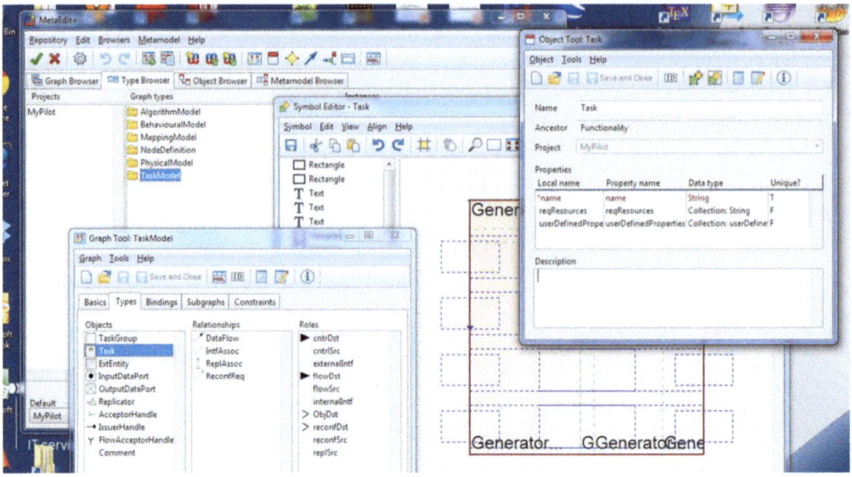

Fig. 3.2 Modeling language specification in MetaEdit+

the meta schema itself) to become reflective: all models can now have explicit references to their object types, as a consequence of which they also become navigable and are seamlessly integrated with the model. Hence, code that navigates the model, can also access and navigate the type definitions and even access and navigate the type's type definitions, etc.

3.2.2.1 Meta Modeling with MetaEdit+

Different meta-modeling tools support (slightly) different meta-schemas. MetaEdit+ [4–6] has a built-in meta-schema which is called GOPPR (Graph-Object-Property-Role-Relation). GOPPR is self-conformant: MetaEdit+ contains a language definition (specified using the GOPPR primitives) that describes the GOPPR schema and can be used to define a modeling languages interactively using the very same environment.

Because graphical modeling languages need to be specified, it is also necessary to define graphical presentations of the objects and relations.

In MetaEdit+, this is accomplished using graphical symbols. Each object type and relation type has a graphical symbol associated with it. Symbols can be composed of other symbols and of (built-in) graphical primitives, like line, arrow, rectangle, etc. The language designer also indicates the graphical artefacts of a symbol to which roles of relations can connect.

MetaEdit+ also provides a built-in language that can be used to navigate the models, generate text and perform checks on the models. Using this language, a code generation function was created and integration with other tools was accomplished by generating language specific XSD files and for each model an associated XMI

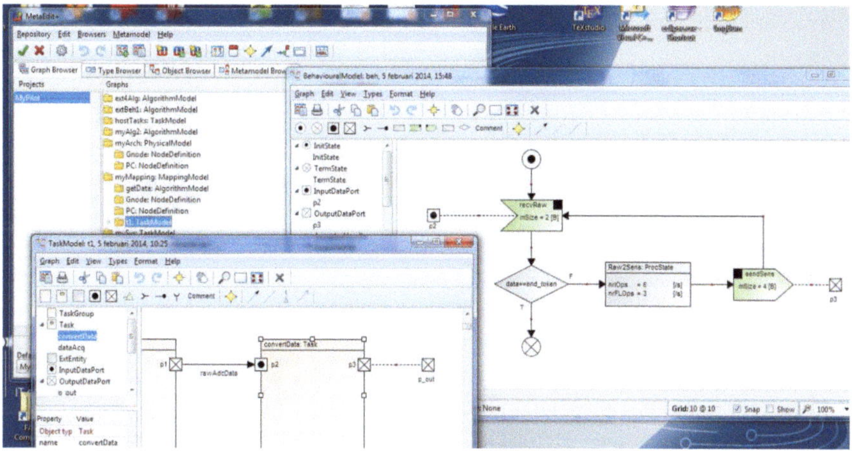

Fig. 3.3 Model editor (tool) created using MetaEdit+

file. The XSD and XMI files are used as an intermediate format for exchanging the models among the different tools of the tool chain.

To provide support for a user-defined modeling language using a meta-modeling tool like MetaEdit+, the following activities need to be carried out:

- Define a graph type, specifying a new language (or aspect of a language)
- Define all object types and relation types in that graph type
- Define the properties of all object types
- Define the roles of all relation types
- Defining the graphical symbols for representing and manipulating the objects and relations
- Specify modeling language constraints (in a declarative manner)
- Implement generators to verify more complex language constraints and to support code/report generation

For a more detailed tutorial and modeling examples, we refer to the MetaEdit+ tutorial and extensive documentation at http://www.metacase.com/.

Figure 3.3 shows an example of a graphical modeler in MetaEdit+, supporting the language defined in Sect. 1.2.

3.3 Quantitative Evaluation and Optimization of System Designs

The second group of activities in the design phase comprehends the quantitative and qualitative assessment of the system: verification of its properties, feasibility, and constraints, and the refinement of the system design. These activities are supported by a large variety of tools, including simulators, formal verifiers, analyzers, optimizers, and model transformation tools. This section focuses on tools for the quantitative assessment and optimization of the design.

The design of large, distributed and reconfigurable embedded systems becomes often a difficult challenge due to the size of the design space. In fact, the number of variables in the design phase—such as allocation of resources and functionalities, resource parameters, and structural organization of the components (network architecture)—frequently end up in an explosion of design choices. This results in complex, emerging, and unpredictable system level behavior, which means that there is no obvious relation to small changes in some parts of the design and the overall behavior of the system.

For such systems, the traditional way of coping with failures that are part of nominal operation, in for instance large scale sensor and actuator networks, is the introduction of redundant elements. The disadvantages are obvious: the system is more expensive and power hungry, not adaptive to unknown situations, and hence still a risk of complete system failure is present. The alternative is to design a system that is fault tolerant (or capable of graceful degradation), which implies that the system can continue performing its tasks, but at a reduced level. Measuring the performance for these type of systems is not trivial. In this section, we describe the approach is implemented in the analysis and design tool called DynAA [7].

Designing an embedded system implies that there is freedom in various design parameters. These design parameters involve three distinct elements in the system:

1. Functional or application specifics software and algorithmic choices.
2. Physical or architectural characteristics number of components and component types.
3. Assigning the applications to the architecture.

All combinations of parameter choices lead to a large so-called design space, and each point in this space represents a complete instance of a system. If all points in the design space are evaluated, the best system is known and hence the best design can be selected. This is however more often than not an unfeasible task. Therefore, an efficient exploration of the design space is necessary. In a general way and as depicted in Fig. 3.4, the exploration of the design space presents three phases:

Initialization defines a measure that captures the performance of the design instances;

Exploration as long as no termination criterion is met:

(a) Select a system design instance,
(b) obtain the behavior of this instance, and
(c) measure the performance of the instance.

Termination select the best performing design instance.

Such phases are discussed in more detail in the remainder of this section.

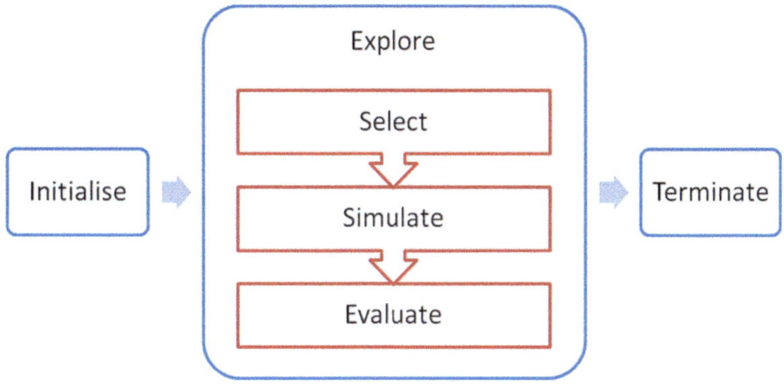

Fig. 3.4 Design space exploration

3.3.1 Modeling for Design Evaluation

In order to be able to measure the behavior of some design, the designed system and its environment are modeled, after which a simulation is carried out. The system model consists of four model blocks, which combined cover the variation in the design parameters and the interaction with the environment. Each model consists of small elements that can be described relatively easily and in a unified manner.

Functional model describing the algorithms and tasks and interactions.
Physical model hardware components and their interactions.
Allocation model mapping of functional to physical elements.
Execution model interaction with the embedding environment.

The functional model involves elements such as OSI layer protocols and tasks that process sensor measurements. The physical model describes hardware components such as processors, batteries, memory modules and communication devices. The execution model describes when sensor measurements are available and emergence or mobility of physical elements, but can also introduce of failures in communication, break down of hardware components, re-allocation of tasks.

This means that all elements in the system design are modeled, after which a simulation engine carries out the actual simulation. During the simulation, we keep track of the relevant properties for a proper design evaluation.

3.3.2 Design Evaluation

A system design is successful, if the system implements some predefined desired functionality, while meeting certain system level requirements. However, the design that is of interest is not one that only satisfies all the constraints, but also performs better than other designs in terms of additional properties. Some of these properties or characteristics can be system level, others are more local. Well known system properties that give an indication of performance include power consumption, throughput, message loss, total memory usage, processor activity, response times and system availability. The characteristics that are relevant for evaluating the designed system are defined in the initialization phase and are known as key performance indicators (KPIs) and each KPI is a measure for the designer to evaluate a design and hence enables the comparison of different designs.

Before the simulation is started, monitoring entities are added to the system model, which log the KPI values throughout simulation. In this way, a good understanding of the behavior of the design is obtained. After a simulation, the designer obtains the KPI values and can interpret them to value the selected design. In case of an automated design space exploration loop, the interpretation should be provided by the designer beforehand. This is described in more detail in the next section.

3.3.3 Input for Design Exploration

For an automatic design space exploration, the designer should provide input for the following three mechanisms:

- a selection mechanism,
- an evaluation mechanism, and
- an exploration mechanism.

3.3.3.1 The Selection Mechanism

The designer provides a complete system model, which includes a (large) part of the system that is fixed in each of the designs, and the dynamic part, which is the design space that is to be explored in order to select the best design. One way of providing this model is by applying the code generation tool to the XMI files that are generated based on the Demanes modeling language.

The designer can—if needed—crop the design space by providing setup constraints, which means that certain points (hence designs) in the design space are unfeasible and should be excluded. The following simple example can clarify this principle: suppose that a certain sensor node restricts the protocols that can be selected for routing, simple because they are not implemented on that node. For other nodes, these protocols are implemented and hence part of design space. This means that there is a relation between the parameter that describes the selection of a sensor node and the parameter that describes the routing protocol. This relations are described in the setup constraints.

The type of constraints here involves constraints on the design of the model itself, that is, constraints that are based on model parameters. For the evaluation of these functions the actual construction and simulation of a model is not needed, which means that these constraints are dealt with separately. A reasonable approach within design exploration is to first determine a system instance that satisfies the setup constraints, after which the simulation is carried out. (However, different approaches on exploration and optimization exist.)

3.3.3.2 The Evaluation Mechanism

The evaluation mechanism involves an interpretation of the KPIs, which is a mapping of KPI values to a single cost value by the so-called cost function, and the designer should provide a function in such a way, that a smaller cost value indicates a better design. This approach enables the designer to provide an evaluation mechanism that is tailored for the application. The evaluation mechanism is of great importance for the result, and it can be very difficult to define a proper cost function that maps all relevant KPIs of a complete system design to a single number, in such a way that the smallest cost implies the best design.

The setup constraint functions ensure that the design is feasible in terms of design parameters, but also performance constraint can be present. To put it differently, a successful design is needed, which means that is satisfies all requirements. Examples of such constraints are: the response time of the system can never exceed a certain maximum time, or: the consumed power in some piece of hardware cannot exceed the total capacity of the battery in that hardware component. In short, these constraints have at least one or more KPIs as input and hence need a system model simulation before the constraint function can be evaluated to see whether the current setup is sufficient.

These constraints that need a design simulation for evaluation can be treated by adding a punishing cost to the cost function in such a way, that a larger violation involves a larger punishment. By choosing the punishing cost properly, it can be shown that the constraint violation does not exceed some small value (known as tolerance). A mathematical formulation of this approach is given later in this section.

To provide additional flexibility, the performance constraints and cost function can also accept properties of entities as input parameters. In this way, the cost function can, based on a certain set of properties, behave differently. For instance, it can be the case that the designer wants to assign different cost values for a KPI on message loss, based on the selected routing protocol: if a protocol that includes acknowledgments is selected, the message loss can be less relevant.

3.3.3.3 The Exploration Mechanism

The exploration mechanism is a way of traversing through the design space. Since the designer is interested in the best performing design, constraint to a set of system requirements, it is obvious to consider an optimization strategy as mechanism. Selecting the right optimizer is strongly related to the type of the problem (both in a mathematical and in a domain specific sense), the size of the design space and the computation time that is needed for a simulation run. By allowing the designer to provide the optimizer, the characteristics of the problem can be exploited.

There are two difficult aspects in any system design optimization problem that we mention explicitly. The first is the presence of integer constraints. These constraints result from the translation of a finite set of design choices for parameters in the design space to the mathematical equivalent of a set of natural number. In addition to that, the relaxation of these integer constraints, combined with for instance a cut, branch and bound strategy cannot be applied. The reason is that we cannot map the resulting non-integer values in the relaxed problem to a proper parameter value in the model directly. Only in the case of a small number of integer constraints with only a few parameter values, a tree of parameter choices can be constructed for the evaluation, but this approach cannot be applied to problems with more than just a few integer constraints.

The second aspect involves the computation time of a single simulation and evaluation round. Since this time is relatively large, the number of evaluations (and simulations) during the optimization process should be minimized. A genetic algorithm for instance, needs a very large number of function evaluations for convergence and is therefore not advisable.

Optimization algorithms have been designed to specifically deal with these kind of hard problems. An example is the use of a radial basis function that is constructed during the design space exploration, which is used to approximate the cost function. The actual optimization is done alternately evaluating the radial basis function (which is much cheaper in terms of computation time) and the actual cost function (which needs a full evaluation and simulation round).

3.3.4 Models for Optimization

Exploring (traversing) the design space in a relatively efficient way is key to determine a good (or maybe even the best) design within a reasonable amount of time. This means that a translation of design choices is needed to a vector of numbers, such that each point in the mathematical space corresponds uniquely to a point in the design space and hence a unique system design.

The canonical form in which mathematical optimization problem are written is

$$\begin{array}{ll} \min & f(\mathbf{x}), \\ \text{subject to} & g_i(\mathbf{x}) \leq 0,\ i = 1, \ldots, n, \\ & \mathbf{x} \in \mathscr{C}. \end{array}$$

The constraints that are mentioned here, involve the setup constraint functions. The performance constraints $h_j(\mathbf{x}) \leq 0,\ j = 1, \ldots, m$, are evaluated after simulation and combined with the designer-provided cost function $f(\mathbf{x})$, as follows:

$$f(\mathbf{x}) = J(\mathbf{x}) + \sum_j w_j \max[0, h_j(\mathbf{x})].$$

Here, the values w_i are punishing values, and their value should exceed the magnitude of the objective function to force feasibility to these constraints.

To put the above mathematical formulation into a context of system design, we state an example for some sensor network that estimates some physical attribute by filtering and fusion of sensor measurements.

$J(\mathbf{x})$ the average accuracy of the computed estimate of some physical parameter,
$h_1(\mathbf{x}) \leq 0$ the total energy consumption in the sensor nodes can never exceed the capacity,
$g_1(\mathbf{x}) \leq 0$ if the selected communication device is NONE, the fusion should be NONE as well since that needs communication (would like a better example here..),
$\mathbf{x} \in \mathscr{C}$ there is one out of a total of four fusion algorithms selected in the processing unit (that is, the involved variable can only attain a value in the set $\{0, 1, 2, 3\}$).

The exploration mechanism and optimization methods presented above allows the designer to use domain specific knowledge and hence is able to provide a tailored set of KPIs and evaluation mechanism. In addition, the optimization approach is selected.

This information that is provided by the designer is parsed and the optimization loop is started. At each loop iteration, each parameter is assigned one of the possible values (hence just a point in the design space) and the corresponding system model is constructed. After construction, the model is simulated, which results in a set of target key performance indicators (KPIs). These values are fed into the objective function, which return the evaluation back to the optimizer. Based on the results, the optimizer proposes a new model parametrization and as long as no stopping criterion is met, a new iteration starts: new choices for the parameters are determined by the optimizer and a new model is constructed, simulated and evaluated.

3.3.5 DynAA

DynAA is a tool that implements the model construction and simulation steps. DynAA comprises of

DynAA core a discrete event simulator,
DynAA model generic modeling concepts in form of a library of basic components,

DynAA library a library of application independent component models,
DynAA user library application specific model components.

3.3.5.1 Core

The core of DynAA is a simulation environment that models the operation of a system instance. This system instance can for instance be the result of the code generation on the XMI file (see Sect. 3.1).

The simulation core has two primitives: entities (any instance of a model component) and events (an observable occurrence within the simulation). Each event occurs at a particular instant in time and marks a change of state of one or more entities in the system. An event is created (scheduled) by some entity, and zero or more entities in turn respond to that event (by providing a callback function while registering to events that are yet to happen).

An event has a source (the entity that generates the event), a type and an ID. The type enables entities to distinguish between the nature of events. The ID or event identifier uniquely identifies the occurrence of the event in the complete simulation.

The simulation of the embedded system in its environment is performed by traversing over time chronologically. This means that the events that occur on the current point in time are processed and only if no events are left on that time, the simulator advances discretely in time to the next time-instance on which events occur. In between events, no changes in the system (are assumed to) occur. We note that new events can be created on the current point in time, which occurs if entities that are processing the current events are scheduling new events with a delay of zero units of time.

3.3.5.2 Generic Models

The generic model components describe common elements in any (embedded) system. The generic functional elements are described in tasks, while the physical reality is described in nodes, which are the physical platforms on which the tasks are executed, and the environment, in which all nodes live. Note that for tasks such as user defined algorithms we do not need to actually run a real implementation of the algorithm that performs/executes a particular task (during simulation). Instead, a model that reflects global features on its execution (describing interactions, sending/receiving/storing data, computational demands) is used.

For tasks that are related to the OSI-layer, the actual protocols are followed to simulate the behavior properly.

The environment is responsible for maintaining a set of services. Examples include a list of where nodes are located, and providing sensors with measurements. The environment is a placeholder for the execution model. Examples of interactions: availability and location of nodes or other hardware components, the transfer of information, and sensor measurements. One particular role for the environment is that is a placeholder for scenario-dependent events.

A node is a bundle of related peripherals, processors, communication devices, and memory and battery modules. Generic models of node, which consists of and peripherals such as a battery (which contains of a default discharge model that describes the evolution of capacity and voltage), and a sensor (which can measure some quantity in the environment) communication device.

3.3.5.3 The Model Library

The model library contains domain specific primitives for building distributed networked systems and are extensions (or specifications) of the generic model components. The components in the library enable a quick modeling of a system. We describe a few of the components in the model library.

With regard to functional part of the embedded system, most components are related to communication. For instance, a generic routing protocol framework is available, but also various widely used routing protocols have been implemented. In addition to functionality on the network layer, we implemented also protocols on the data link and physical layer. For instance, the CSMA/CA (Carrier Sense Multiple Access met Collision Avoidance) protocol has been implemented. This protocol acts on the data link layer of the OSI-model, and is a network multiple access method in which carrier sensing is used. It is an extension of the aforementioned generic OSI-Layer component, which is part of the generic model.

For the physical part of the model, various components exist. For instance, an implementation of the GNode (a wireless sensor node) is available, and the CC1101, the transceiver on this node (an extension of the generic communication device component), is implemented as well.

A batteryStateLogger is a logger that is registered to change events of battery properties and in this way logs the battery's capacity and voltage along the simulation time. A more generic logger is the propertyChangeLogger.

Additionally, various useful services that are registered to the environment component are available in this library. Examples include the node location map (describe location of each node) and line of sight map (describe for each node whether some other node is visible). These two methods can be used for instance to compute what the signal to noise ratio of messages that are send from one node to the other.

3.3.5.4 Own Components

If some physical or functional component is case-specific or for other reasons not available in the DynAA library, the designer has the possibility is possible to create his/her own components. The generic or library components can serve as a basis or inspiration for the construction of these components.

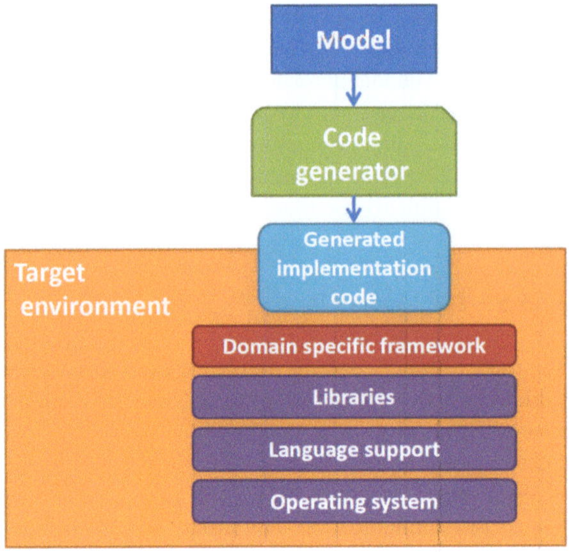

Fig. 3.5 From model to implementation code. The generation of implementation code from highly abstract models benefit from the existence of a domain specific framework. Such framework captures part of the model semantic and helps bridging the abstraction gap between model and programming language

3.4 Runtime Services

A model-based approach is not only suitable for the design time activities discussed in the previous Sections. It also helps and accelerates the implementation of the system, when the model can be automatically *transformed* into software implementations. Such task is possible but not always easy. Such transformation can be a difficult challenge due to the gap in the level of abstraction between the model and the implementation code. On one hand, the system model is typically described in a high level of abstraction and using constructs (languages) with much hidden semantic concepts. On the other hand, implementations tend to be extremely detailed with aspects that are not welcome in the model, such as hardware specific configurations, etc.

In order to ease this step and allow a more efficient conversion of the model information into implementation code, one possible strategy consists in increasing the level of abstraction on the target platform. That can be done with a pre-defined 'middleware' layer—typically a set of functional libraries or an application framework—that captures the most important aspects of the modeling languages semantics. If such strategy is used, the process receives the model as input and produces an executable code tailored to a certain target environment, as depicted in Fig. 3.5.

For the target systems we aim for in this book, the domain specific framework and the libraries on which they are based can be seen as *a hardware and application independent software layer that provides functions, algorithms, and design patterns typically found on adaptive systems*. The function of this middleware layer is to support, capture, and integrate common implementation practice of distributed **a**daptive, **n**etworked, and **e**mbedded systems. It other words, it contains standard design pattern implementations for reconfiguration mechanisms, information dissemination and aggregation mechanisms, and management of adaptive software components. Therefore,we will call this software layer the **ANES middleware** throughout this chapter. The use of an ANES middleware will have the following advantages over implementing the functionality directly from the application into a programming language:

- The design and implementation of the application specific knowledge becomes independent of the implementation of the adaptation mechanism underlying it. The designer concentrates on solving the problem of its application domain, and uses the runtime environment as a library or service to carry out the modifications in the system.
- The designer can reuse functional components that are common and typical to many adaptive systems. Moreover, it creates a base for interoperability between different applications as it defines the underlying mechanisms used for adaptation. As an example it defines how data is disseminated in a network, or yet how new components are inserted and removed from the system.
- The runtime environment allows including, excluding, or modifying components of a system or application during its execution time in a standard way. Also, it allows the adaptation mechanism to be exchanged without modifying the application using it. As a result, part of the problem of designing adaptive systems can be postponed to the execution time, instead of being planned in advance during the system design.
- Mechanisms for system adaptation relies on basic, known building blocks (functions, services) that are easier to be validated and analyzed when isolated from the application specific details. Such building blocks also make it easier to describe and analyze the expected behaviour of the application where they are used.

In the following we discuss the functionalities that are desired in an ANES middleware. We divide these functionalities based on the aspect of implementation they represent, and we call these aspects **runtime services** because they constitute in large a series of standard software modules and services that will be available during the application runtime for supporting adaptation. The 'must have' runtime services of an ANES middleware are:

- Support for a runtime system composition through reconfiguration and module lifecycle management;
- Support for managing the adaptation process;
- Support for resource monitoring;
- Support the construction for system service or distributed systems;

3.4.1 Support for a Runtime System Composition Through Reconfiguration and Module Lifecycle Management

The first important runtime service of an ANES middleware is supporting dynamic composition of a system through reconfiguration. Managing software components at runtime is at the core of an adaptative system. Runtime services have to offer to application designers the possibility to add, remove, replace and group software components at runtime without having to shutdown or reboot the system:

Adding Components Although application structure is usually well-thought by designers, it often necessary to be able to add software components at runtime without having to reboot the system due to the fact that safety critical requirements demand for an always-on type of execution. An ANES middleware should offer an API for adding software components at runtime based on user-defined interface (input signals, output signals, module byte-code).

Removing Components When a certain software block is not useful anymore, the framework should allow removing this software component to liberate resources and 'simplify' the system.

Replacing Components Consists in replacing a given software component byte-code. Such a service is necessary in case of recurring software bugs, detection of security vulnerabilities, requirements changes, etc.

Assigning Tasks At the most basic level, the execution of a software component has to be assigned to a given operating system (virtual machine) task. Software management should offer a function call to perform such an action either based on user input or in an automated manner.

By using a software component management system, coupled with an extensible library of software components, arbitrary applications can be implemented on a Networked Embedded System at runtime. If the software components are of the right abstraction, and implement generic enough functionalities, they can be wired, rewired and parameterized in such a way that very specific functionality can be obtained, either on the level of individual devices, or at the network level. Extensibility of the software component library allows for new hardware (for instance sensors) that introduce new application opportunities as well as new communication channels, to be leveraged within applications, so as to conform to evolving application user requirements.

3.4.2 Support for Managing the Adaptation Process

In a general way, it can be said that any adaptive system follows a model similar to the one depicted in Fig. 3.6. Part of the adaptivity behaviour is related to collecting information about the system (observers), part of it is related to reasoning about the information and detecting conditions that trigger the adaptation (reasoner), and part

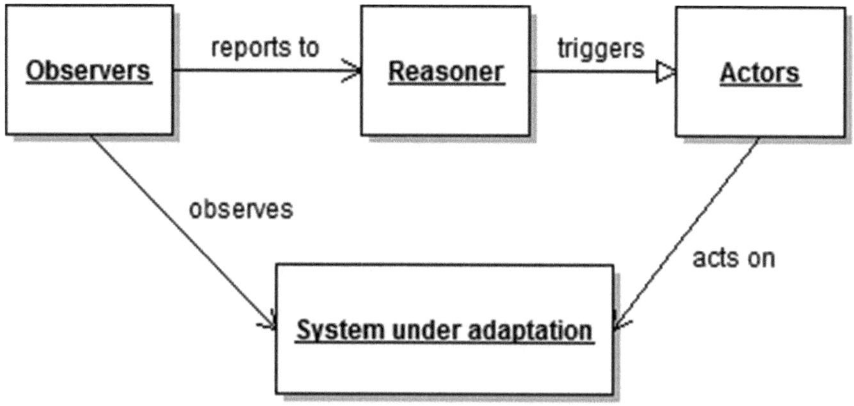

Fig. 3.6 Generic model of an adaptive system. Observation of the environment, reasoning about the actions to take, and acting on the system are typical activities of an adaptive system

of it is related to executing a certain adaptation strategy (actors). As an ANES runtime service, we identify as management of adaptivity the techniques and procedures involved and interfacing with decisions (reasoner). In this sense, an ANES middleware must provide a generic and flexible interface that helps application developers to plug and play their specific reasoners.

In this sense, an ANES middleware should provide:

- A standard interface that reasoner modules can use to collect information from observer modules pull method. Also a standard interface that observer modules can use to push information into reasoned modules push method;
- A standard interface to register, activate, and control reasoner modules;
- A standard interface for reasoners to trigger actuation modules;

The implementation of a reasoner module can follow different methodologies, for example ontologies, first-order rules, fuzzy logic, machine learning, etc. However, in general the reasoner works as a rule engine, where human knowledge can be encoded as rules or formulas. Those rules are the policies the system can use to take decisions about when to configure and the best configuration of services and resources. Besides, reasoners and their rules have many practical benefits, such as flexibility to change them dynamically avoiding to re-compile the application and its logic, or the restrictions defined in rule engines that add consistency to the whole reasoning system.

3.4.3 Support for Adaptive Networking and Communication

ANES systems relies on seamless communication, automatic sensor transformation and data aggregation. Sensors used in ANES system shall be adaptive in variety of environments and circumstances in order to be used for many purposes where applicable. Seamless communication consists on capability to join network automatically without prior knowledge of devices, establish a connection on application level and adapt to location of a user. In addition, data communication shall be designed in a way that different applications and sensors can communicate each other using list of available services and data fields or similar method. In case of a node failure occurs, the network should reconfigure itself in order to sustain operation for the rest of the network and avoid service downtimes. The base of the home automation relies on positioning a user. Thus when a user moves rerouting occurs, user goes out of the sensor range and inside the measurement range of some other sensors. To avoid unnecessarily network load, ANES system will also detect when a user can be serviced using WLAN/LAN or 3G. Similar sensors can be used in variety of different locations. To establish derived measures, the data coming from a sensor must be able to be delivered for several application points.

3.4.4 Support for Resource Monitoring

Resource monitoring can be considered as obtaining information concerning the utilization of one or more system resources (CPU power, bandwidth, memory, storage, etc.). As the systems providing high critical services cannot fail or low perform, they shall use software and hardware agents to reduce or avoid situations of potential failures. This means that sensors and monitors (HW/SW) provide information about events occurrences related to a specific situation of failure or low performances. This information is shared with other systems that are in charge to monitor the situation and make decision for problem solving. In this context, the information collected by the monitoring system support the adaptive system making it able to check the situation and modify the behaviour of other components in order to maintain the critical functionalities correctly active.

Resource Monitoring in ANES ecosystem relates to a variety of devices such as servers, measurement devices, sensors and mobile devices. Many of the measurement devices and mobile devices are coming with rechargeable batteries. Data storage of devices is usually limited. So data and available energy is limited resource in some devices of ANES ecosystem. This introduces a need for energy consumption modeling and data storage monitoring. Sensors and other similar devices can have also self-diagnostics available. ANES system should monitor connected devices in order to alarm users for faulty devices. This helps to keep on quality of measurement data, as data of possibly bad sensor can be eliminated.

3.4.5 Support for Service-Oriented Component Architecture

Distributed processing management requires to enable discovery and access to available services and resources in the environment in order to use the offered functionalities and capacities at the best way. Specifically ANES systems envision an environment where services and resources provided by devices, systems and third-parties are seamlessly able to connect, discover and share relevant data. In this way, the ANES middleware will enable evolution inherently so the final system is able to easily adapt and reconfigure.

For doing so, ANES should provide a publish and subscribe mechanism to facilitate the service and resource discovery. It should grant the communication among devices that use different communication protocols (such as TCP/IP, Zigbee, Bluetooth, Konnex) and that were programmed using different programming languages (Java, C/C++, Android, Python). A semantic approach would reduce the heterogeneous data issue, granting each element in the system is using a common vocabulary. The capability to check sensors status and consistency processing capability needs the set-up of seamless integration and management of WSNs. This means using the Web Ontology Language (OWL) where the specific sensors domain concepts will be modeled and represented. A common Ontology has to be provided that will be able to homogenise all the incoming data so it can be stored in a structured form making use of the semantic web technologies like RDF (Resource Description Language). Therefore REST APIs are needed that embed semantic web query languages like SPARQL to query the stored RDFs sensors data to get sensors measures, set actuators states, check the correct functioning of the WSNs and to detect inconsistencies.

3.5 Conclusions

This chapter proposes a basic composition for a model-oriented tool chain. The composed tool chain covers the most important activities of the design phase, namely model editing, model visualization and transformation; model validation and evaluation; and (semi-) automatic system implementation based on the system model. The chapter also discusses how to prepare the model for design space exploration.

An short survey on embedded systems modeling tools is presented. Extensions for the modeling language and modeling techniques are presented that cope with system reconfigurability and (dynamic) scalability of the system.

References

1. A. Ledeczi, M. Maroti, A. Bakay, G. Karsai, in *Proceedings Workshop on Intelligent Signal Processing* (2001)
2. R.C. Gronback, *Eclipse Modeling Project: A Domain-Specific Language (DSL) Toolkit*. The Eclipse Series (Pearson Education Inc., Boston, 2009)

3. S.G.H. Dubois, F. Lakhal, in *MARK'09 Proceedings of the 2009 Second International Workshop on Managing Requirements Knowledge* (2009), pp. 85–88
4. V.T.K. Smolander, K. Lyytinen, P. Marttiin, in *CAiSE'91 Proceedings of the Third International Conference on Advanced Information Systems Engineering* (1991), pp. 168–193
5. K.L.S. Kelly, M. Rossi, in *Proceedings of CAiSE'96, 8th International Conference on Advanced Information Systems Engineering*. Lecture Notes in Computer Science, vol. 1080 (Springer-Verlag, Berlin, 1996), pp. 1–21
6. M. Rossi, J.P. Tolvanen, in *Proceedings OOPSLA'03 Companion of the 18th Annual ACM SIGPLAN Conference on Object-Oriented Programming, Systems, Languages, and Applications* (2003), pp. 92–93
7. Z.Papp, J. Oliveira, Model-based design of self-adapting networked signal processing systems in *Proceedings of the International Conference on Self-Adaptive and Self-Organizing systems conference (SASO)* (2013)

Chapter 4
Runtime Validation Framework

Roshan Kotian, Stefano Galzarano, Claudio Bacchiani, Aly A. Syed, Přemysl Šucha, Roman Václavík and Andrei Pruteanu

Abstract Testing large-scale complex systems at runtime is of paramount importance. This is particularly true for dynamical systems, such as distributed adaptive network embedded systems (ANES), which exhibit adaptive capabilities aiming at autonomously reconfiguring and adjusting their behavior based on the changing environmental conditions. In such cases, it is not feasible, during the development stages, to anticipate all the possible operating conditions that the system may face in a real environment. This is because some information about the execution context and the system itself can be available only once the system has been deployed. Thus, in order to correctly assess the effectiveness, efficiency and robustness of ANES, it is required to verify that the system correctly adopts the proper adaptation mechanisms in response to the context changes as well as to check the quality of such adaptations. The focus of the chapter is to discuss about the needs for employing runtime verification and validation of ANES and the main challenges and requirements for its implementation. In addition, it presents a reference framework that

R. Kotian (✉) · S. Galzarano · C. Bacchiani
TU/e, Den Dolech 2, Eindhoven, The Netherlands
e-mail: r.kotian@tue.nl

S. Galzarano
e-mail: s.galzarano@tue.nl

C. Bacchiani
e-mail: c.bacchiani@tue.nl

A.A. Syed
NXP, High Tech Campus 60, Eindhoven, The Netherlands
e-mail: aly.syed@nxp.com

P. Šucha · R. Václavík
Czech Technical University, Prague, Czech Republic
e-mail: suchap@fel.cvut.cz

R. Václavík
e-mail: vaclarom@fel.cvut.cz

A. Pruteanu
TU Delft, Postbus 5, Delft, The Netherlands
e-mail: a.s.pruteanu@tudelft.nl

Z. Papp and G. Exarchakos (eds.), *Runtime Reconfiguration in Networked Embedded Systems*, Internet of Things, DOI 10.1007/978-981-10-0715-6_4

supports developers in testing adaptive systems at runtime. One of its key feature is the capability to emulate certain realistic conditions through synthetic data, which is useful to check the system's behavior under specific and controlled situations.

4.1 Introduction

System testing is an integral part of software engineering and involves testing an entire system based on its requirement specification. It includes activities such as functional and non-functional testing [1]. Although there exist many system testing methodologies for traditional software systems, verifying and validating large-scale adaptive network embedded systems (ANES) requires more development. Since ANES such as Wireless Sensor Network (WSN) modifies its behaviour according to the changes in the environment, testing them poses new set of challenges [2]. The success of ANES depends on its ability to adjust its operation in real-time due to external factors and/or internal factors, and therefore verifying whether the system has done adaptation as per its design requirements is of great importance. Failure to perform verification of adaptation might result in ANES producing inferior quality of service. Apart from verifying that the ANES has adapted correctly due to the changes in the environment or due to its own system state, it is also important to check the quality of the adaptation to boost the overall efficiency of ANES. However, lack of well-defined metrics to assess the adaptation qualities of the system makes the vision of achieving runtime verification and validation hard [3]. Therefore, it is important to devise a runtime verification and validation framework that not only tests the adaptation, but also incorporates certain metrics to assess the adaptation quality of the ANES. Before the ANES goes live, it would be quite handy to have a sub-system within the testing framework that emulates certain realistic conditions through synthetic data. This type of integrity checking helps the ANES to be more robust and reduces the failure rate post deployment. Other pressing issue related to testing of the ANES are to extract the requirements from the user and convert them into necessary test-cases. In this chapter, therefore we briefly elaborate our framework that addresses the key issue mentioned above. We first highlight the need of employing runtime testing for ANES; the challenges related to the implementation of runtime testing for ANES. We then outline the requirements of our framework for ANES. We conclude the chapter by presenting the general overview of our testing framework.

4.2 Needs for Runtime Verification and Validation in ANES

To test the working of traditional software one can apply design-time testing such as model checking [4] because they have fixed execution conditions that are elucidated during software requirement phase. In addition, the operational environment where traditional software is deployed remains unchanged. On the other hand, the main objective of ANES is to provide uninterrupted service even in the changing environmental conditions. Hence, it may have several paths of execution. Typically, it is difficult to anticipate all possible operating conditions that an ANES may face throughout execution during development stage [5]. ANES requires rigorous runtime verification and validation because it may be subjected to conditions for which they were not explicitly designed [6]. In ANES such as WSN, it is often the case that some information from the system is available only at runtime. For example, checking if the adaptation has happened or evaluating the quality of adaptation can be done only when the system is executing. It is not feasible to identify at design-time the entire possible operational context that an adaptive system may encounter at runtime [7]. ANES such as WSN are often deployed in a remote, hostile environment such as battle field; water bodies etc. This exposes the system to prolonged adverse condition during which the system has high probability to deviate from its initial design. To mitigate this, we need the system to perform runtime verification and validation the moment the system model drifts from its original goals. Runtime verification and validation greatly reduces the human intervention to repair the network.

4.3 Challenges of Runtime Verification and Validation in ANES

In the previous section, we put forward the importance of having runtime testing. In this section, we look in to some of the main challenges of implementing runtime verification and validation techniques for ANES. Large scale adaptive systems contain many numbers of individual autonomous embedded systems that collectively performs certain operation. This distributed nature of operation makes it difficult to implement a testing framework that verifies and validates large scale adaptive systems. For example, consider a node in a large scale Wireless Sensor Network (WSN) adapting their transmission power level at runtime due to degradation in the propagation medium. A node by increasing its transmission power guarantees that the data reaches its neighbor with higher reliability. However, its effect in terms of interference, latency is more global. Testing the effects of increasing the transmission power level at a global level is often difficult because of the sheer size of the network. All nodes that are affected by the local operation must share the information with another and a system testing entity must be executing in all the resource constraint nodes. An ANES that re-configures or adapts itself as specified by the functional requirements partially fulfils its objectives. To continue having good quality adaptation, the system must also satisfy non-functional requirements. An example of non-functional requirements would be how stable and consistent the adaptation is. Absence of standardized metrics to evaluate the non-functional requirements of ANES makes the task of implementing runtime Verification and Validation (V&V) cumbersome. Therefore, integrating quality metrics from control theory and re-interpreting it for developing runtime testing framework is imperative [3]. Providing feedback on quality of adaptation to the system is valuable as it enhances the reconfiguration mechanism. In order to achieve this, it is necessary to address the not so straight forward task of numerically quantifying the quality of adaptation which is more subjective in nature. Testing traditional software systems are easy as they have fixed set of inputs and expected outputs. ANES are designed with initial set of requirements and assumptions. However, these prior set of specifications may no longer be valid as ANES operate in non-deterministic ever-changing environmental conditions. As a result, pre deployment design of the adaptive system may fail causing frequent redesign and maintenance. To avoid costly maintenance, verifying the resilience of ANES to failure is the desired option. However, devising a sub-system within V&V framework that emulates all possible realistic worst case scenarios remains one of the key fundamental challenges.

We know that the initial design of the ANES must be verified and validated at runtime. However, for this to be realized there must be a provision made in the testing framework for a sub-system that extracts the test-cases from the requirement specification. A test-case generation sub-system aids the system designer to instantly validate the initial design of ANES. Automatic generation of test-cases is often non-trivial task as it requires technique to extract appropriate keywords from the requirement specification and then translate them to executables test-cases. Since techniques such as compositional verification, program synthesis, model checking provides more reliable way of checking the goodness of the software at design-time, investigating the mechanism of implementing these features at runtime is an important challenge [3].

4.4 Runtime V&V Requirements for ANES

We addressed the need for runtime testing in previous sections. In this section, we provide the key requirements that are necessary to implement a runtime testing framework for ANES. The main objectives of runtime testing framework are to verify the functional and the non-functional properties of ANES. A functional requirement specifies a function that a system must be able to perform what the product must do. One of the main requirements of ANES is to be in its viability zone. Viability zone can be defined as the set of system states where the systems requirements and desired properties are not compromised. By executing the self-adaptation mechanisms a reconfiguring procedure planned and executed by the ANES, it maintains itself in a viable state [3]. ANES operates in an uncertain environment. Hence, the viability zone of ANES conceived during the development stage based on certain assumptions may drift when the network is deployed. Therefore, the main requirement of runtime testing framework is to check if the systems viability zone has drifted from the one that was designed during the design stage. The feedback from the testing framework aids the system designer to build the model that closely reflects the operational condition prevailing at the location where the system is to be deployed. Non-functional

Table 4.1 Key Performance Indicators for adaptive networked embedded systems

Aspects of ANES	Key Performance Indicators
Functional	Adaptation reliability, latency, stability, check if the system is within the specified viability zone
Non-functional	Settling time, consistency, resource utilization for adaptation

requirements describe the non-behavioral aspects of a system, capturing the properties and constraints under which a system must operate [8]. It addresses the quality of adaptation performed by ANES. To have robust ANES it is not enough to verify the adaptation performed. It is equally important for example to know how fast and how much resource was consumed in the process. The functional and non-functional properties or requirement of ANES are expressed by Key Performance Indicator (KPI) and Service Level Objectives (SLO). KPI is a particular measure representing the performance of running system. For example, response time, accuracy etc., are valid KPI that provides valuable insight about the performance of ANES. SLOs are defined to objectively represent functional and non-functional requirements in terms of possible value ranges of tangible and quantifiable measures. Each SLO is expressed as a *<KPI, relational operator, value>* tuple.

The key requirement of runtime testing framework is to identify and verify appropriate KPIs and its corresponding SLOs. Next, we derive the KPI from [3] that can be used to gauge the functional and non-functional aspects of ANES (Table 4.1).

The Adaptation reliability is the property saying the system is able to work correctly under the circumstances for a specific period of time. One can use many metrics to be able say whether the system is reliable enough or not, e.g. defect density, defect intensity, mean time between failures, etc. The accuracy can be described in terms of viability zones. One can be interested not only in the information whether the system is in some viability zone but also in what is the gap between measured values and boundary values defined in each viability zone. For example, if the tests show that the system reaches the maximal desired response time regularly, it is the sign that one should do something with that before the system ends out of any viability zone. Stability is closely connected to functional and non-functional requirements, i.e. if those requirements are not fulfilled the system will not work properly and may cause some significant problems or even damages. On the other hand, if all requirements are satisfied the system works correctly and as was intended. Basically, the system is stable if it is in some viability zone. Settling time and Consistency refers to how fast the system adapts to the desired viability zone and how often adaptation is done given a specific adaptation scenario. The resource usage is the value of the CPU utilization, transmission power utilized by the transceiver of some device. Again, it can be measured quite easily on the assumption that an operating system of device allow user to obtain this value. This KPI measures how much resource was utilized than normally required while performing adaptation.

One of the goals of a good testing framework is to provide a feedback on the behaviour of the software. To this extent, our framework also provides the in-situ mechanism to test the design of the adaptive system by injecting appropriate synthetic data to the system and collecting information on how the system adapted under various emulated scenarios.

Another main functionality of our framework is to capture stakeholders expectations, translate and relate those expectations to adaptation requirements and generate necessary test cases based on the SLO mapped to KPI that tests the adaptation functionality and performance of the large-scale system at runtime.

4.5 The V&V Reference Framework: An Overview

As already discussed in the previous sections, the difficulties in carrying out a set of tests on an adaptive system after its deployment lead to the need for a proper tools capable of effectively supporting the developers through all the required steps involved in the verification and validation process of the system's behavior at runtime.

In this section, we provide an overview of the proposed Verification and Validation (V&V) Framework aiming at facilitating the runtime testing phase. In particular, this framework aims at: (1) checking if the current state of the running system is in a Viability Zone (*System-level Testing*), and (2) assessing the correctness and the quality of the adaptation mechanisms employed by the system (*Adaptation-level Testing*).

In order to accomplish its tasks, the framework has been conceived to perform the following main testing strategies:

- *Testing the system under its normal operating execution mode*: the system is checked as it is, and the test does not influence any condition/properties of the system whose execution only depends on the "normal"interaction with the environment and on the consequent inner state evolution.
- *Testing the system under some induced conditions*: the test involves the injection of "fake" data into the system so that the induced conditions will influence its behavior. Such a capability is useful when the system, which may be already deployed in its operating environment, needs to be tested under specific and controlled situations. For instance, it may be useful to check the adaptive capabilities of the system when specific conditions tend to take it outside the current viability zone.

To assist the Test Designer in effectively testing the adaptive system, the framework supports the following processes: (1) the formalization of the necessary test-cases to be applied to the target system, and (2) the actual execution of the V&V procedure by interacting with the real system under testing. In Fig. 4.1, the overall scheme of the proposed testing framework is depicted.

As it can be seen, the V&V Framework includes the following components:

- *Test-case generator*. It is conceived to guide the test designer in properly defining the test-cases, which are then translated in a well-defined machine readable format.

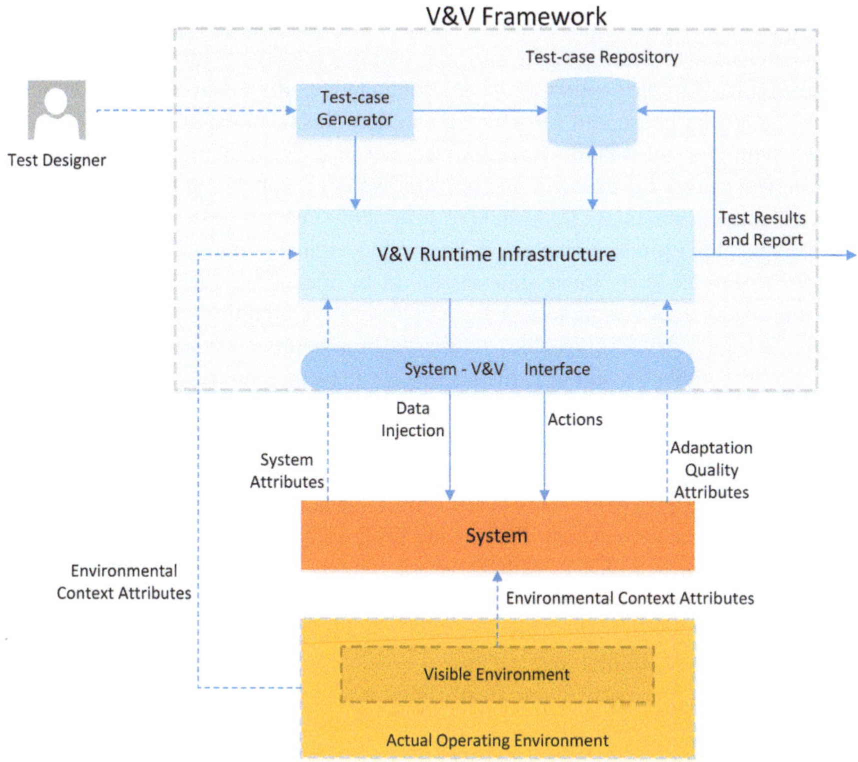

Fig. 4.1 The verification and validation framework

- *Test-cases repository*. It represents a database for storing the test-case descriptors along with their related test results and reports for further off-line analysis.
- *V&V Runtime Infrastructure*. It is the actual executor of the test. Depending on what is described in a specific test-case, this component is in charge of:
  - configuring the runtime testing environment;
  - carrying on all the required V&V tasks by interacting with both the system under test and the environment where the system has been deployed;
  - gathering all the necessary information related to the state evolution of the system and the changes in the environment as a consequence of their interaction but also as a consequences of an active intervention of the V&V Runtime Infrastructure on the sytem.
  - generating reports based on the obtained testing results.
- *System-V&V Interface*. It represents a collection of components specifically designed to allow the interaction between the V&V framework and the target system.

As it can be seen in Fig. 4.1, the importance of a framework is to test the system running in its real Operating Environment is that it is possible to monitor not only the attributes related to the portion of the environment actually taken into consideration during the design phase and for which the system has been explicitly conceived to interact with (i.e. the Visible Environment) but also the remaining portion of the environment in order to assess if and how the system is capable to properly adapt to unforeseen environmental circumstances generating unexpected conditions.

Moreover, the proposed framework has been designed to interact with the system in order to influence its inner state evolution by means of direct actions aiming at tuning the system parameters or even choosing the adapting mechanisms to be applied. In such a way, it is possible to compare the adapting behavior usually adopted by the system under certain circumstances with the one induced by the framework at runtime and assess which one leads to better results.

The system can also be influenced indirectly by changing the environment status. However, since it could not always be possible to really change its real conditions, the V&V Runtime Infrastructure is able to override the actual environmental attributes, which the System senses from the Visible Environment, with some induced (not real) information. In Fig. 4.1, such an interaction is depicted as "Data Injection". Thanks to that, the proposed V&V Framework is also capable of running what we call "*Playback Feature*", which will be described in details in Sect. 4.6.3. By injecting specific environmental and also internal system attributes, which may have been previously recorded or synthetically built, it is possible to carry out tests on the system running over a well-defined setting.

As shown in Fig. 4.2, the main purpose of the System is to implement a set of Primary Functionalities (PF) which transforms a set of input data stream into a set of output stream according to both functional and non-functional requirements. The Reconfiguration Functionalities (RF) are in charge of making runtime changes in the Primary Functionalities by selecting algorithm alternatives, modifying parameters, adjusting tasks allocation or schedule of execution. The V&V Framework interacts with both the set of PF and the set of RF with the capability of interfering in the reconfiguration process at runtime. In order to do that, a set of components, the Monitors, need to be deployed along with the System for gathering all necessary information related to both the PF (i.e. the System Attributes) and the RF (i.e. the Adaptation Quality Attributes). A set of Actuators is also necessary to actually make the proper changes as a consequence of the analysis of the gathered attributes. In particular, as already mentioned, there are two different kind of "actuation", the one directly changing the System by interacting and altering the behaviors of both PF and RF (depicted in Fig. 4.2 as "Actions") and the other dealing with the injection of not-real (environmental or internal) data for creating ad-hoc conditions to influence the System.

In general, the V&V Framework is connected to the System block to accomplish the following tasks:

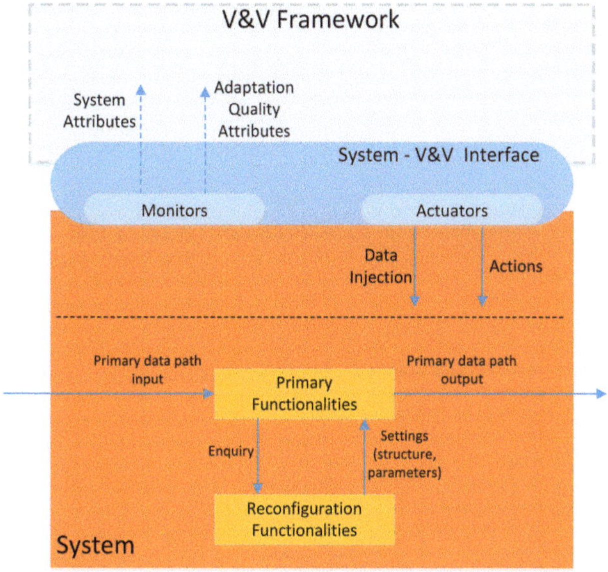

Fig. 4.2 Interaction between V&V framework and primary/reconfiguration functions

- Checking the normal operation of the system as a consequence of the execution of the Primary Functionalities, which represent the main purpose for which the system has been conceived.
- In the case that a Reconfiguration Functionality has been executed on the system, verifying if the reconfiguration action has been successfully performed by verifying if the system is still in a Viability Zone.
- Evaluating a metric to measure the quality of reconfiguration, e.g. accuracy, settling time, consistency and resource usage of the reconfiguration process.

4.6 The V&V Runtime Infrastructure

This section describes in more details the components constituting the framework and in particular the ones in charge of actually executing the tests on the system.

Once the test designer has properly formalized (with the help of the Test-case Generator) a specific test-case to be applied to the target system, the corresponding machine-readable description is provided to the V&V Runtime Infrastructure.

As shown in Fig. 4.3, the main components constituting the V&V Runtime Infrastructure are the following:

- *Test Controller*: it represents the orchestrator of the entire V&V Runtime Architecture.

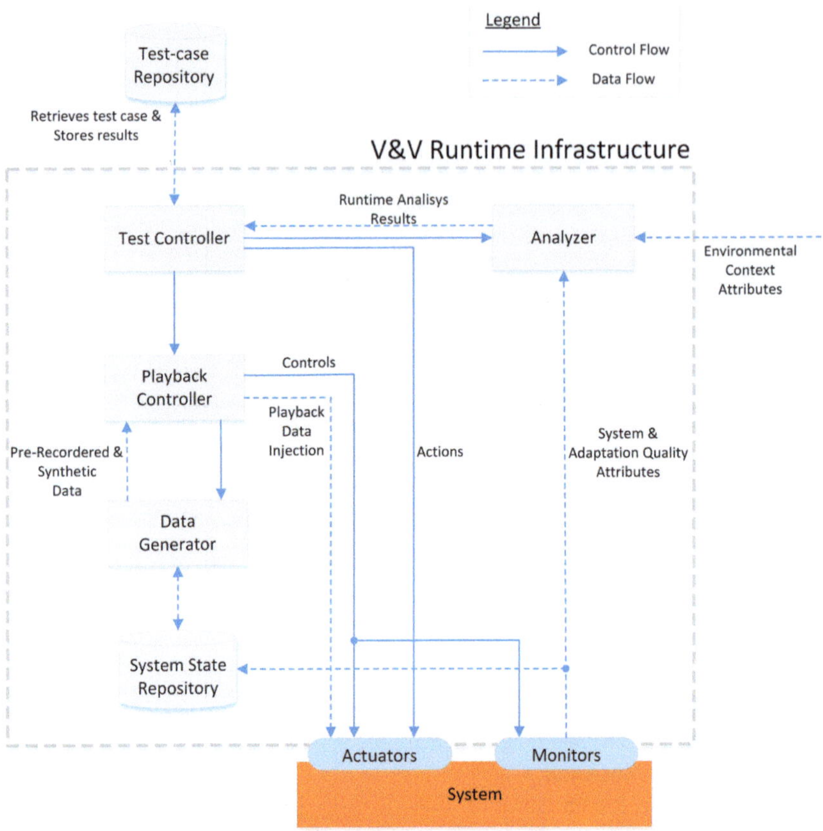

Fig. 4.3 The verification and validation runtime infrastructure

- *Analyzer*: it incorporates the necessary algorithms and techniques for carrying out two main tasks:
  - checking if the system is in a Viability Zone;
  - assessing the quality of the system's adaptation mechanisms;
- *Playback Controller*: it is in charge of managing the playback feature of the framework.
- *Data Generator*: it supports the Playback Controller by generating proper data streams to be injected into the system.
- *System State Repository*: it is a storage component designed to keep track of the system's states evolution to be provided to the Data Generator.
- *Monitors* and *Actuators*: represent the interfaces between the framework runtime components and the actual system and are in charge of gathering information from it and having an effect on the system respectively.

The Test Controller is in charge of interpreting the description of the test-case and configuring and controlling the other runtime components accordingly. In particular, it directly configures the Analyzer by setting up the proper analysis algorithms that best fit with the desired testing requirements, and by selecting which aspect and attributes of the system should be monitored. Specifically, it instructs the Analyzer to access the needed subset of attributes provided by the Monitors. Monitors are specifically designed to be system-dependent since they requires to be deployed along with the system or even to be directly plugged into it in order to probe any useful attributes related to the current system state, the context in which the system knows it is deployed (i.e. the Visible Environment depicted in Fig. 4.1) and the adaptation functions adopted at runtime. They can also be implemented to retrieve some performance statistics to help the Analyzer to easily and quickly carry out its analysis. However, as already discussed, along with the information coming from the Monitors, the Analyzer also needs further attributes related to the actual environmental context in which the system is deployed but of which it is unaware (i.e. the Actual Operating Environmental depicted in Fig. 4.1). This is of fundamental importance to assess how the system reacts to unknown and unpredictable changing environmental conditions.

Once the system is deployed and the test is running, the Test Controller continuously interacts with the Analyzer to check the outcomes of the analysis and, in case the system does not behaves as expected, it will directly alter its normal behavior by means of the Actuators, which are expressly conceived to tune system's parameters, modify its execution control and, in general, change any other settings both in the Primary Functionalities and in the Reconfiguration Functionalities. Similarly to the Monitors, these interface components are also system-dependent.

The Analyzer is triggered by the Monitors in response to two types of events. These events are generated during the post deployment life time of the system either when an adaption function is executed, or as a consequence of regular maintenance monitoring aiming at checking the whole or parts of the system. The Analyzer then tests if the system is able to keep itself in a viability zone or not. In the case when an adaptation function is performed, the Analyzer also evaluates the reconfiguration action based on the adaptation metrics reported by the Monitors.

Significant components of the V&V Runtime Infrastructure are the ones related to the playback feature. In particular, the Playback Controller is responsible of controlling the injection of fake data into the system in order to create specific conditions into the system so as to evaluate its behavior under a well defined environment and situations. In particular, there are two types of fake data that can be provided by the Data Generator: pre-recorded and synthetic. The first one is directly taken from the real system's state history whose information has been acquired by the Monitors and stored into the System State Repository. Conversely, synthetic data streams are ad-hoc generated based on what specified in the test-case description or possibly also built as a modification of the system states stored in the repository.

4.6.1 System Monitoring

The main function of the monitoring components is to provide information about the state of the system in order the V&V runtime sub-system to determine the viability zone and reconfigure the system when it leaves a viability zone to bring it back into one. The monitored information is also necessary to determine the quality of adaptation functions at runtime according to the criteria specified by the Test Designer. Therefore the data provided by the Monitors can be directly fed to the Analyzer module or stored in the System State Repository for reproducing the same conditions via the playback function of the verification and validation infrastructure.

Thus, the monitors handle three types of information:

- Environmental: information that describes the context in which the system operates. This information can be about the sensors in or around the system that observe the environment. An example could be a temperature sensor that senses the environment temperature that may affect system function.
- Internal: describes the state of the system before and after the reconfiguration, including internal parameters and the outputs. An example of this could be amount of memory consumed at a particular time in a computing system that could change its behavior.
- Adaptation: when the reconfiguration has occurred, these monitors can provide information regarding the executed adaptation functions that can help determining the quality of the adaptation process. For instance, it can be the time that it took to reconfigure the system, or the amount of resources used for the reconfiguration.

Monitoring can be periodic, on-demand, event-driven or a combination of these. Periodic monitoring occurs when a system variables is observed at fixed time intervals, on-demand requires a command from the Test Controller or the system itself, whereas event-driven monitoring is triggered when a pre-determinate condition is verified (e.g. temperature raises beyond a threshold). It is important to preserve the temporal correlation of the recordings from different sources (i.e. environmental and internal data), especially when causality between different inputs must be preserved. This also holds when recording data, where it is necessary to ensure that the correct sequence of events is reproducible.

4.6.2 System Analysis

The Analyzer component is an important part of the verification and validation framework. It serves to decide whether the tested system works properly under given conditions and circumstances or not. The Analyzer has to consider two main criteria to be able to determine the correct functionality and behavior of the tested system:

- Viability zones: whether the tested system is in one of the allowed VZs.
- Adaptation quality: certain quality level of the adaptation process has to be reached.

Fig. 4.4 The system-level V&V workflow

Such analysis needs, for its successful work, interactions with other components of the V&V runtime infrastructure, namely the Test Controller and the system Monitors. The Test Controller triggers the Analyzer for a particular test case. Depending on it, the controller may also trigger the Playback Controller to inject some pre-recorded data or synthetic data into the system. The Analyzer then verifies the expected functionality and behavior derived from the particular operational conditions against the actual output from the real system provided by the Monitors. Finally, the result of that comparison is returned to the Test Controller.

In particular, two different roles in the V&V framework can be distinguished, namely a system-level V&V, where the framework determines if the system is in a viability zone, and adaptation-level V&V, where it determines the quality of adaptation. Each role has slightly different workflow from the Analyzer point of view.

The system-level V&V workflow is depicted in Fig. 4.4. It starts with gathering quantifiable system and/or environmental variables, so called attributes. These attributes basically describe the current state of the system and its execution context. From these real measures taken at runtime, the Analyzer is able to build a well-defined system representations. Then, it is necessary to determine all possible viability zones with respect to the measures which are derived from the functional requirements of the observed system. Specifically, each viability zone is formalized as a n-dimensional space and defines a subset of possible system states in which the system operation is not compromised, i.e. it is working properly and all conditions together with requirements are satisfied. The task of the Analyzer is to determine whether the system representations (which are also defined in terms of n-dimensional spaces) are in some of the allowed viability zones. Consequently, the Analyzer notifies the Test Controller with the results.

Figure 4.5 represents the adaptation-level V&V workflow for the analysis related to the chosen adaptation functions, which is quite similar to the previous workflow. At the beginning of the whole process some quality attributes have to be calculated from a set of proper measurements. These attributes speak much about the quality of the adopted mechanisms to determine how good or bad the adaptation was. Finally, the measured quality attributes, such as a latency and reliability, are also sent to the Test Controller, which is in charge of taking proper actions.

Fig. 4.5 The adaptation-level V&V workflow

4.6.3 Playback Feature

As already previously discussed, the role of the playback feature is to inject particular data (pre-recorded or synthetic) into the system for verification purposes. The expected behavior is compared against the actual behavior of the system when known conditions are (re)played back. This helps in the identification of bugs/failures/inconsistent behavior. As part of the V&V tool, the playback feature aims at facilitating the:

- Mapping of certain conditions to the appropriate VZ. These conditions might be the current or past ones, synthetic or both. The conditions may be classified as *normal*, if within VZ, or *abnormal*, if outside.
- Quality of the adaptation under a given condition.

In order to perform the playback feature, the Playback Controller interacts with special monitoring components called Virtual Monitors to control the injection of this fake data streams, as depicted in Fig. 4.6.

When the playback feature is adopted during a specific test case some of (or all) the real monitors needs to be overridden by their corresponding virtual monitors. A Virtual Monitor is in charge of intercepting the values delivered by its corresponding real Monitor. These virtual monitors have to be placed as close to the real monitors as possible so that the response of the system to those replayed conditions can be monitored at all levels. The virtual monitors should be able to monitor all the three modes of the system: environmental, internal and adaptation. Each virtual monitor

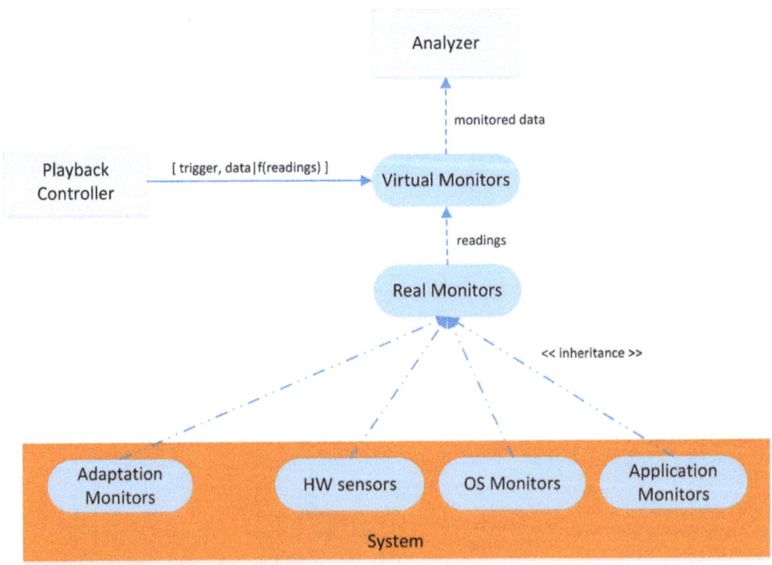

Fig. 4.6 The virtual monitors

is basically a piece of software or hardware directly mapped on top of a real one, and through which it is possible to filter the real data by providing "fake"generated one. Specifically, different types of real monitors are present in the system depending on what they monitor:

- System-level monitors: provide measures about the status of the hardware, operating system and applications.
- Adaptation-level monitors: are in charge of providing the adaptation quality attributes.

Once monitored data or synthetic data need to be injected into the system, the Playback Controller configures each virtual component by means of a specific setup procedure represented by the *<trigger, data|f(readings)>* tuple:

- *trigger*: a flag meaning that the virtual monitor has to be activated.
- *data*: represents data (single data or an entire data stream) that has to be injected into the system. It is extracted either from the System State Repository containing the recorded states of the system or synthetically built by the Data Generator.
- *f(readings)*: instead of explicitly inject stream of data into the system, a virtual monitor may be configured to be a more complex filter by setting a certain function able to directly manipulates the data readings coming from the real monitor.

There are three main "data relations" in the diagram depicted in Fig. 4.6. The following list describes those options in the most general way. Moreover, the basic interfaces are introduced from the virtual monitor point of view:

- Real Monitor → Virtual Monitor: the output/readings of the specific real monitor goes directly to the related virtual monitor since the real data has to be modified by the injected data/function as close to the real monitor as possible. Provided interface:

  – *getReadings(KeyValuePair readings):void*—where key is a unique name of the measured quantity and value can be either primitive data type or the complex one depending on what one wants to extract from the monitor. For example, in case of temperature measuring, the key could be string "temperature"and value could be integer "25".

- Playback Controller → Virtual Monitor: the playback controller can activate or deactivate any virtual monitor to be able to control the output of real monitors through the injection very precisely. The injection is done either by direct data or by function taking into account even the real readings which can be beneficial in some cases. For example, in case of temperature measuring, it would be possible to emulate an addition of some heating device by incrementing the real environmental temperature. Provided interfaces:

  – *getInjection(KeyValuePair data):void*—the description of the parameter is the same as for the interface *getReadings*.

  - *getInjection(Func<KeyValuePair, KeyValuePair> fnc):void*—where *fnc* is a function manipulating the input data with one parameter of type *KeyValuePair* and the return value also of type *KeyValuePair*.
  - *getActivity(Boolean active):void*—where the parameter active determines whether the virtual monitor is active or not.

- Virtual Monitor → Analyzer: finally, the readings have to be sent from the virtual monitor to the analyzer. Two cases can occur: (1) the virtual monitor is deactivated so the readings from the real monitor are passed unmodified to the analyzer; (2) the virtual monitor is activated so the readings are modified in some way and after that they are passed to the Analyzer. Required interface:
  - *setReadings(KeyValuePair monitored_data):void*—the description of the parameter is the same as for the interface *getReadings*.

As for the synthetic data, there exist many ways to generate data streams. All methods have one important thing in common and that is a system model. The synthetic data can be only created with the knowledge of the system. Otherwise, the data would not make any sense if it did not correspond to the system model which describes the real system under test. The easiest approach is stochastic generation of all data. However, the resulting quality is not satisfying and it is really hard to control the coverage of the above mention situations. Moreover, some data can be even infeasible, i.e. completely non-sense combination of values. One can create handmade data. This is quite precise technique and fully controlled by the knowledgeable human. Unfortunately, it is very time demanding. Simulations can serve to generate data efficiently supposing the model is well designed and corresponds to the real system. The rule-base approach is able to capture even complex relations between measured data and, thus, it creates more sophisticated data. Bayesian models are also used for the accurate data generation. Nevertheless, it also requires a solid knowledge about statistics from the designer.

4.7 Testing Workflow Examples

After having described in the previous sections the purposes and the functionalities of the components constituting the proposed T&V framework, this section presents a description of the individual steps that need to be carried out in order to complete the activities involved in the T&V process. The main aim is to guide the reader in understanding the whole testing operations through well-defined sequences of tasks. In the whole validation and verification workflow, two main (and subsequent) sub-workflows can be identified:

- Test-case definition;
- Test-case execution.

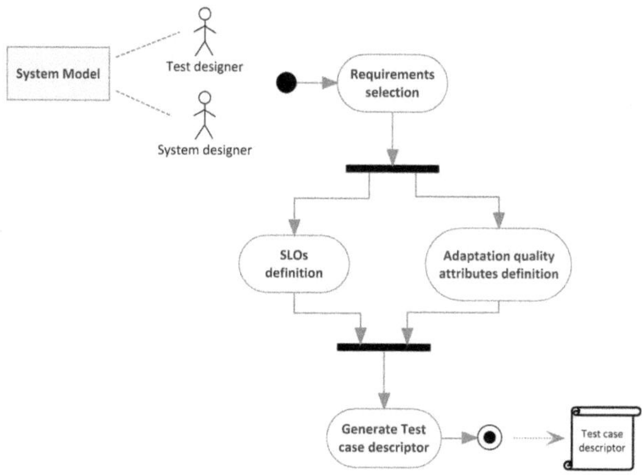

Fig. 4.7 Test-case definition

The test-case definition process directly involves both the System Designer and the Test Designer which, depending on the specific System Model, are in charge of detailing all the information required to instruct the testing runtime system to properly execute the test. The final output of this process is the definition of a machine-readable configuration file (the Test-case descriptor), which is fed to the interpreted by the Test Controller (see Fig. 4.3). Based on what kind of testing strategies the test designer is interested in, we can have two possible different tests-case:

- Testing the system under normal operating execution mode: no playback functionality;
- Testing the system under some induced conditions, i.e. by enabling the playback functionality.

No matter what strategy is employed, the steps involved in the Test-case definition can be represented by the workflow of Fig. 4.7.

The initial step is a very important task because all the following ones heavily depend on the selected requirements (functional and non-functional) related to which the system is going to be tested. Based on the selected requirements, the designers are then in charge of defining both a set of SLOs and a set of adaptation quality attributes. These tasks can be carried out in parallel because they refer to two distinct testing aspects, as discussed in Sect. 5.4. The SLOs determine which metrics have to be considered during the test for checking the system behavior with respect to its Viability Zones (System-level testing), whereas the adaptation quality attributes are considered for assessing the system adaptation capabilities. All these evaluation metrics are then aggregated and stored in a machine-readable configuration file, which represents a formal description of a specific test case. According to the main purposes of the T&V framework, once its components are deployed along with the system under testing, different flows of tasks may occur at runtime, depending on the specific event that activates the testing architecture:

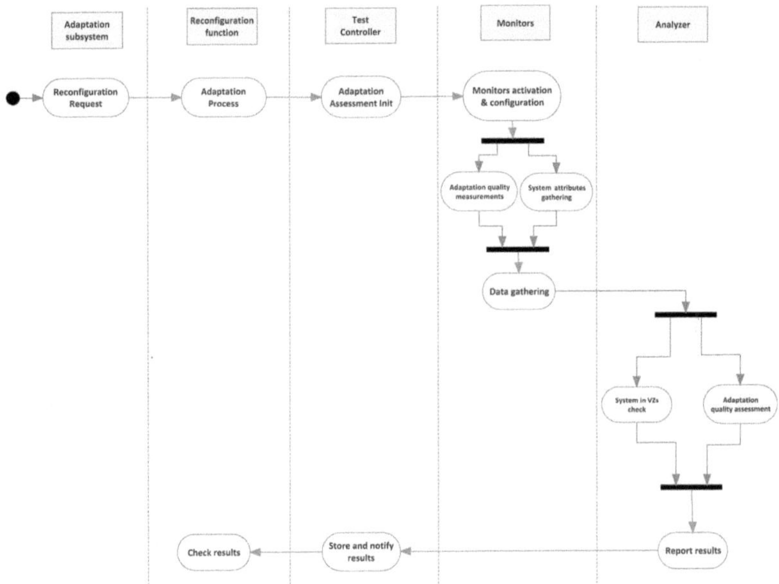

Fig. 4.8 Adaptation assessment workflow

- As soon as a reconfiguration operation is required in the system and a proper adaptation process is initiated, the T&V architecture is immediately triggered to perform the necessary tasks in order to both properly evaluate the quality of the adaptation under execution and to check if such an adaptation process actually keeps the system in one of its Viability Zones.
- Test Designer may also define some maintenance operations which are scheduled in order to test the running system under specific circumstances. In this case, the Playback functionality of the T&V architecture is usually triggered by the Test Controller and employed to create certain conditions different from the real ones.

The activities performed upon the occurrence of an adaptation process can be effectively described by the workflow represented in Fig. 4.8.

The whole process is initiated when some reconfiguration activity occurs in the system. As a consequence, a specific adaptation process is started by the system to properly manage such a request and, at the same time, the Test Controller is triggered to initiate an adaptation assessment procedure. One of the main important task to be executed is the activation and configuration of a set of Monitors, which are in charge of:

- measuring all the necessary quality attributes related to the execution of the adaptation;
- measuring the context attributes describing the current state of the system.

Once all this information has been gathered, it is provided to the Analyzer component, which is triggered in order to analyze it by means of specific techniques capable of:

- assessing the quality of the adaptation under execution;
- determining the current system state and checking whether it is still in one of its Viability Zones.

The resulting outcomes are aggregated and sent to the Test Controller. The Test Controller may send this information to the Reconfiguration function. In the case when the Test and Validation system is being used at design time, then the system designer may be informed about the results. In the case that a system maintenance is being carried out, then the Test Controller informs the maintenance personal about the result of Test and Validation. Different from Fig. 4.8, the workflow representing the activities of the testing framework as a consequence of a specific scheduled maintenance operation involving the Playback functionality is depicted in Fig. 4.9.

In this case, the test execution procedure is triggered by some scheduled plan defined in the Test case descriptor and set into the Test Controller by means of an initialization task at deploy-time. Such maintenance plans may be defined to periodically assess the correct execution of the system under certain well-specified conditions. Thus, since the Playback component plays an important role, the Test Controller needs to provide it with proper information related to which, when and how data has to be injected into the system at runtime. As a consequence of this configuration phase, the Playback Controller first sets and activates all the necessary Virtual Monitors deployed in the system and then manages the actual dissemination

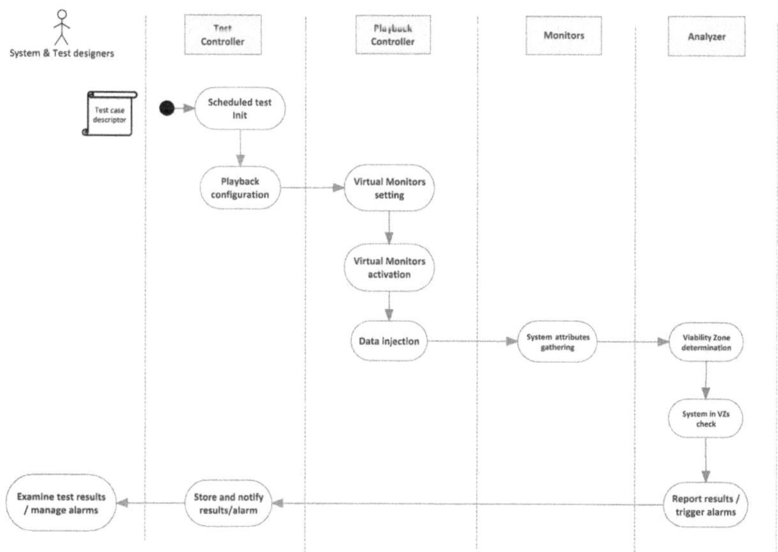

Fig. 4.9 Scheduled maintenance system workflow

of the data for creating "non-real" environmental and state conditions. The Context Monitors can then monitor the evolving context attributes so to allow the Analyzer to properly analyze the response of the system to those conditions. In particular, the same basic tasks of Fig. 4.8 are executed: based on the current context attributes of the system, the Analyzer is able to determine whether the system adaptation processes at runtime are keeping the system in one of its Viability Zone. If not, a proper alarm is immediately triggered. All the results computed by the Analyzer are then aggregated and sent to the Test Controller, which in turn notifies the System and Test designers of the identified bugs/failures/inconsistent behaviors in order to proceed with the necessary procedures.

4.8 Conclusions

In this chapter, we have looked into the details of having runtime testing for adaptive systems. We, highlighted the challenges associated with the implementation of runtime testing framework. Concentrating on the needs and challenges, we provided the general reference framework for our proposed runtime verification and validation framework.

In the consequent chapters, we provide the detailed explanation of our proposed runtime framework.

References

1. L. Briand, Y. Labiche, Softw. Syst. Model. **1**(1), 10 (2002). doi:10.1007/s10270-002-0004-8, http://link.springer.com/10.1007/s10270-002-0004-8
2. N. Petschenik, *System Testing With an Attitude* (Dorset House Publishing Co., Inc., 2005). http://www.dorsethouse.com/books/sta.html
3. R. Lemos, H. Giese, H.A. Müller, M. Shaw (eds.), *Software Engineering for Self-Adaptive Systems II*. Lecture Notes in Computer Science, vol. 7475 (Springer, Berlin, 2013). doi:10.1007/978-3-642-35813-5, http://www.springerlink.com/index/10.1007/978-3-642-35813-5
4. C. Baier, J.P. Katoen, *Principles of Model Checking (Representation and Mind Series)* (The MIT Press, 2008)
5. E.M. Fredericks, B. DeVries, B.H.C. Cheng, in *Proceedings of the 9th International Symposium on Software Engineering for Adaptive and Self-Managing Systems, SEAMS 2014* (ACM, New York, NY, USA, 2014), pp. 17–26. doi:10.1145/2593929.2593937, http://doi.acm.org/10.1145/2593929.2593937
6. J. Zhang, *A Formal Approach to Providing Assurance to Dynamically Adaptive Software*, Ph.D. thesis, (East Lansing, MI, USA, 2007). AAI3264259
7. E.M. Fredericks, A.J. Ramirez, B.H.C. Cheng, in *Proceedings of the 8th International Symposium on Software Engineering for Adaptive and Self-Managing Systems SEAMS'13* (IEEE Press, Piscataway, NJ, USA, 2013), pp. 169–174. http://dl.acm.org/citation.cfm?id=2487336.2487363
8. M. Glinz, in *15th IEEE International Requirements Engineering Conference (RE 2007)* (IEEE, 2007), pp. 21–26. doi:10.1109/RE.2007.45, http://ieeexplore.ieee.org/lpdocs/epic03/wrapper.htm?arnumber=4384163

Chapter 5
Tools and Methods for Validation and Verification

Paola Jaramillo, Andrei Pruteanu, Willem van Driel, Wijnand van Kooten and Jean-Paul Linnartz

Abstract An intelligent lighting system capable of runtime self-adaption to occupants behaviour is an example of a context-level application in which faulty operation has a strong and undesirable impact on the occupants comfort. Especially in cases where the long-term functioning of the systems is of interest, the systems quality should proof very high and therefore proper validation and verification practices are required. To our knowledge there is non an existent tool that deals with testing runtime self-adaptive systems. In this chapter we propose the implementation of a V&V framework previously introduced, by merging several already known tools. First, we give an understanding of ways to quantify and predict the reliability of large-scale distributed systems. Second, key performance indicators of the self-adaptive systems are identified from monitoring techniques and third, the test cases are formalized in a structured form. We present two test cases as examples of a system working under normal operation conditions as well as under induced conditions, based on real-life implementations. Execution of the test is lead by a test coordinator for which we used JSON notation, and then the interpretation and testing is carried out in a c++ toolbox where the monitoring and testing algorithms reside.

P. Jaramillo (✉) · J.-P. Linnartz
TU/e, Postbus 513, Eindhoven, The Netherlands
e-mail: p.a.jaramillo.garcia@tue.nl

J.-P. Linnartz
e-mail: j.p.linnartz@tue.nl

A. Pruteanu
TU Delft, Postbus 5, Delft, The Netherlands
e-mail: a.s.pruteanu@tudelft.nl

W. van Driel · W. van Kooten
Philips Lighting, Eindhoven, The Netherlands
e-mail: willem.van.driel@philips.com

W. van Kooten
e-mail: wijnand.van.kooten@philips.com

Z. Papp and G. Exarchakos (eds.), *Runtime Reconfiguration in Networked Embedded Systems*, Internet of Things, DOI 10.1007/978-981-10-0715-6_5

5.1 Introduction

As the functionality of computer operations becomes more essential and yet more complicated and critical applications increase in size and complexity, there is a need for elaborate testing. This involves the definition of methods, services and tools for this purpose.

Yet the creation of testing tools, starts with a proper understanding of ways to quantify and predict the reliability of these systems in various complex operating environments.

A (complex) system is a set of interacting or interdependent components forming an integrated whole. When the number of components and their interactions hugely increase, the so-called large or complex systems are formed. The types of components, their quantities, their qualities and the manner in which they are arranged within the system have a direct effect on the system's reliability. The commonly used description for system reliability [1] is given as:

The probability that a system, including all hardware, firm-ware, software, and their interactions will satisfactorily perform the task for which it was designed or intended, for a specified time and in a specified environment.

This definition implies the recognition that most practical systems are not guaranteed error-free.

In fact, we also see approaches to probably-error-free software, for instance in military, automotive or medical equipment. Yet these approaches become prohibitively difficult if the size of the system increases, the operating conditions are not fully predictable, or if components from different suppliers are merged without full disclosure of all contents.

This book chapter discusses the implementation of methods for testing based on the V & V (Validation & Verification) framework for large-scale distributed systems introduced in Chap. 4 [2]. First, the basic concepts involved with software reliability are presented as a primary guide for defining key performance indicators (KPIs) of the system that will consequently lead to the design of testing cases. Following the primacy that a system is structured from several components, KPIs are identified also from monitoring modules. Finally, a set of mechanisms for testing are described in detailed, using as a reference an already implemented distributed *self-adaptive system*.

Initially, the formalization of the test case involves several design steps, as follows. The end-user defines a Software Level Agreement (SLA) in which Service Level Objectives (SLOs) are detailed as functional and non-functional requirements. An interface translates such SLOs into KPIs that later on are aggregated with other requirements in order to create the system Viability Zones (VZs).

Executing the test case requires three main modules: a test coordinator that controls the different components of the V & V framework, a system monitor (s) that reads the environmental context or simulates it from a playback module, and an analyzer that contains the testing algorithms for aggregating the monitored variables and verifying the system status in terms of VZs and adaptivity quality, based on the predefined KPIs.

This chapter describes the implementation of two test cases, when the system operates under induced conditions and when the system operates under normal conditions. Details follow on how to implement a test coordinator using JavaScript Object Notation (JSON) that is later interpreted by the Context Recognition Network Toolbox (CRNT), in which monitoring elements and testing algorithms reside.

5.2 Related Work

Software reliability models can provide quantitative measures of the reliability of software systems during development processes. Research activities in software reliability engineering have been conducted over the past four decades, and many software reliability models have been proposed. A general practice is to determine the system reliability by combining the reliability models of each and all hardware and software components. Even though, much literature is available on the prediction of software reliability [3], and even more is available on the prediction of system reliability based on its hardware [1], little literature is available on the interaction of the above. Alternative models focus on the analysis of system dependability where the system's reliability and availability become integrated as attributes of the system's dependability [4]. All of these previously mentioned models are individually dedicated to test the system at software and hardware levels, thus remains the question of how to translate such reliability requirements into key performance indicators for aggregating them into more extensive testing tools such as the V & V framework proposed in Chap. 4 [2].

Currently, researchers are interested in developing testing solutions for the so-called *self-adaptive systems*, that can support them on overcoming faulty operation modes at runtime, or to simply adapt to new conditions. Recent work from De Lemos, et al. [5] explores the challenges of the design and implementation of *self-adaptive systems*, giving insights into the methods for software validation and verification. However, specific testing mechanisms are required that provide a better understanding of the system's reconfiguration capabilities. Large scale and distributed systems simply add more complexity to the design and implementation of such mechanisms.

A common test method for both the system's components and the overall system is to expose the device to sufficient overstress to bring the time to failure to an acceptable level. This approach is called accelerated life testing (ALT) or overstress testing and it is a form of *Playback Feature* from the V&V framework described in Chap. 4 [2]. A very good reference on ALT was written by Nelson [6]. However, acceleration on a system level is not without risk. Overstress by simply increasing the loads, for example temperature or electrical power, may drive certain components to new and/or unwanted failure modes that influence the actual field performance. Thus, acceleration should be taken with precautions.

Recently, it has been demonstrated that existent platforms such as the Context Recognition Network Toolbox (CRNT) [7], play an important role in the deployment and testing of *self-adaptive systems* [8]. This particular c ++ parameretizable software tool comprises ready-to-use monitoring and testing modules, with the advantage that new functionalities can be easily incorporated. Furthermore, implementations of the CRNT had shown its capabilities for handling distributed signal processing. Although this seems to be a compatible tool to implement the V & V proposed in Chap. 4 [2], still additional components are needed and that is the case of the *Test Coordinator*.

5.3 Translating Key Performance Indicators from Software Reliability and Monitoring Approaches

The functioning of distributed embedded systems poses several challenges with respect to runtime performance. Due to the complexity of both the hardware and software platforms, systems need ways to monitor the quality of each and all their hardware and software components. In the case of DEMANES the system status is monitored and subsequently verified in terms of VZs, based on predefined KPIs. Hence, quality of the system is checked via performance indicators and reliability tests that support systems in their different phases of product development and release by means of different techniques, e.g., software reliability growth models.

Input information obtained via sensor data is crucial for monitoring the outside environment. If sampled values are wrong, e.g. out of boundaries, due to various hardware problems, there is a high risk of not being able to track correctly the dynamics of the outside environment. Since external conditions cannot be predicted a priori and are very difficult to model, we cannot establish predefined thresholds for filtering out data coming from the sensors. To cope with such scenarios we have to use advanced filtering techniques able to remove unexpected values.

Information collected from the sensors is not enough for predicting unexpected system conditions. Monitoring of internal components via either software or hardware must be done via self-monitoring services. Information processing and pattern-matching algorithms detect if certain conditions occur that are outside normal modes of operation. Depending on specific components such as sensor aggregation and processing or accuracy of the reasoners within the DEMANES middleware, the normality detection criteria or verification of the system's VZs, varies as following. For processing of sensor information, specific thresholds for various sensor modalities can filter most of the undesired values. For more advanced modules such as data processors and reasoners, we have to employ custom techniques. For example, if the reasoner is in charge of adapting the runtime execution scheduling of communication, one could detect if KPIs such as *Throughput* are within the desired values.

5.3.1 Software Reliability Concepts

There are many differences between the reliability and testing concepts and techniques of hardware and software. Software reliability or robustness, is the probability of failure-free software operation for a specified period of time and environment. In this sense, software failures are considered a primary cause of product reliability problems, hence a reasonable KPI for testing the software reliability is the *number of software failures* left in the system.

Unlike hardware failures, software systems do not degrade over time unless modified and software failures are not caused by faulty components, wear-out or physical environment stresses such as temperature and vibration. Software failures are caused by latent software defects that were introduced into the software as it was being developed, but were not detected and removed before the software was released to customers. The best approach to achieving higher software reliability is to reduce the likelihood that latent defects are in released software. Unfortunately, even with the most highly-skilled software engineers following industry best practices, the introduction of software defects is inevitable due to the inherent complexities of the software functionality and its execution environment.

A comparison of software and hardware reliability is useful in developing software reliability models. Table 5.1 (input used from [9]) shows the differences and similarities between the two.

Identification and deep knowledge of KPIs at a hardware level is currently driven software engineers to design solutions that are able to reconfigure at runtime, so-called *self-adaptive systems*, in order to overcome some of these hardware failures, and therefore improve the system's reliability. Examples of this can be seen in smart reconfigurable networks, where for instance if certain node stops responding due to lack of power the system is able to identify and activate a new node that replace the non-working sensor including all its functions, so that the system keeps its normal operation and the overall system reliability improves. Testing the software capabilities for runtime adaptation will be discussed in Sect. 5.4.

In conclusion, software reliability relates to errors that are induced by circumstances or contexts that are unforeseen/not addressed in the design phase. In fact, number of software failures left in the system may derive from several causes, e.g., errors, ambiguities, oversights or misinterpretation of the specification that the software is supposed to satisfy, carelessness or incompetence in writing code, inadequate testing, incorrect or unexpected usage of the software or other unforeseen problems. Software reliability is not a function of time, there is not something as software *wear-out*. Hence, typical questions that need to be addressed in order to identify potential KPIs for testing software reliability are shown in Table 5.2.

No method of development can guarantee totally reliable software. A set of statistical modeling techniques are required that, first, enables the achieved reliability to be assessed or predicted, and second, that is based on observation of system failures during system testing and operational use. It uses general reliability theory, but is

Table 5.1 Commonalities and differences between software and hardware reliability

Hardware reliability	Software reliability
Failure rate has a bathtub curve	Without considering program evolution, failure rate is statistically non-increasing
Material deterioration can cause failures even though the system is not used	Failures never occur if the software is not used
Failure data are fitted to some distributions. The selection of the underlying distribution is based on the analysis of failure data and experiences. Emphasis is placed on analyzing failure data	Most models are analytically derived from assumptions. Emphasis is on developing the model, the interpretation of the model assumptions, and the physical meaning of the parameters
Failures are caused by material deterioration, random failures, design errors, misuse, and environment	Failures are caused by incorrect logic, incorrect statements, or incorrect input data. This is similar to design errors of a complex hardware system
Hardware reliability can be improved by better design, better material, applying redundancy and accelerated life testing	Software reliability can be improved by increasing the testing effort and by correcting detected faults
Hardware repairs restore the original condition	Software repairs establish a new piece of software
Hardware failures are usually preceded by warnings	Software failures are rarely preceded by warnings
Hardware might fail due to unforeseen application conditions	Software might (also) fail due to unforeseen application conditions

Table 5.2 Typical questions to be addressed in the identification of KPIs for reliability tests

Observation	Potential KPI
How many failures are left in the system?	*Failures*
In a given time period, what is the probability of no failures?	*Failure probability*
In a given time period, what is the expected number of failures?	*Expected failure*
What is the expected time until the next failure?	*Expected time*

much more than that. In terms of the V & V framework (see Chap. 4 [2]) the following are examples of tests in which the system's VZs are checked (Table 5.3).

Similarly, other types of models are studied, such as the system availability. Availability requirements for consumers are less stringent than for healthcare applications, where availability is defined as:

Table 5.3 System viability zone (VZ) check

Observation	Potential VZ
How reliable is the hardware/software component now?	$80\,\% \leq$ Reliability $\leq 90\,\%$
Based on the current reliability of the software: can we accept it or should we reject it?	Accepted reliability $\geq 95\,\%$
If we accept it, can we stop testing and start shipping?	# Tests ≤ 1000
How reliable will the system be, if we continue testing for some time?	Maximum reliability $\approx 99\,\%$
When will the reliability objective be achieved?	# Tests ≤ 1000
How many failures will occur in the field (and when)?	# Failures at each time t

System availability

The degree to which a system is operational and accessible, when required for use.

In general, a high reliability will always lead to a high availability, but a system with low reliability can still have a very high availability. Reliability and availability belong to the attributes of the system dependability as well as integrity and safety as shown in Fig. 5.1 (input used from [4]). Dependability then is the ability of a system to avoid failures that are more frequent and more severe than acceptable. A dependable system is: having all its required properties, and does not show failures.

5.3.2 *Monitoring Communication Networks*

Self-adaptive systems can automatically mitigate the problem associated with changing conditions of a radio network. In the DEMANES case, such system adaptation is tested via KPIs that can be obtained from the observations or monitoring of the communication network as described in Table 5.4.

The above changes may affect many parameters in the link adaptation, e.g., bit rates and retransmission back-offs. Also it affects higher layers of the protocol, such as network channel changes and routing tables.

Some of the above adaptivity is well covered in communication standards. However increasingly, more rigorous reconfigurations can take place such as a change of roles. For instance changes of which nodes are selected as sector heads and the

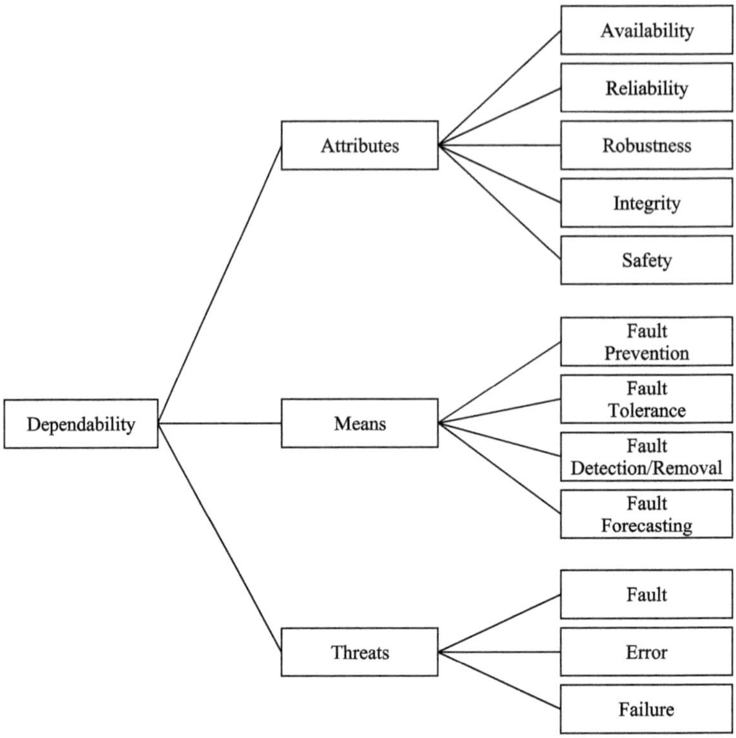

Fig. 5.1 The dependability tree: a measure of system's availability, reliability and maintainability

Table 5.4 Monitoring communication networks to identify KPIs for adaptivity tests

Observation	Potential KPI
Mobility of the users changes link budgets and changes the degree of interference that other links cause	*Channel interference*
Arrival of new nodes	*Network topology*
Nodes that cease to operate within the networks, for instance because of lack of power	*Node status*
Changes in the interference patterns	*Interference*
Changes in the traffic density, for instance because of new demands from applications and services	*Traffic density*

choose of the communication protocol across standards, or within a standard, such as transitions from broadcast to unicast. In the next paragraphs we describe some indicators for congestion or potential KPIs.

- ***Number of retries***: In multiple access networks, a notorious problem is that all nodes in the network need to make many retries when sending a message. These retries lead to more interference such that all other nodes also have to retransmit messages frequently. In turn, this leads to instabilities of the network. A well known example is the ALOHA network.
- ***Back-off***: Many wireless systems adapt the retransmission back-off period depending on the number of messages collisions that it sees. The value of the backoff can thus be used as a canary for system health.
- ***Average channel occupation***: As with many queuing systems, the delays and performance typically stay within a comfortable range as long as the utilization is sufficiently small. On the other hand, the delays typically increase above any finite limit if one tries to load the system up to its maximum capacity. Hence, the average channel occupation can indicate how close the system is to a collapse.
- ***Connectivity***: In wireless networks the number of neighbors that can be heard and that can be reached is a clear indicator of the density of the nodes. Presumably the routing tables and their variability over time can be used to monitor system health.
- ***Stability of routing***: Many wireless sensor networks have the ability to change the frequency channel if too much interference is experienced. If a network starts a channel change too frequently this is a clear indicator of a potential problem.

5.3.3 Monitoring the Application Context

First of all, a system–under–test itself makes imperfect observations of its environment, and different parts of the system may have different interpretations of the contextual reality. In the DEMANES case, a runtime reconfigurable system that is subject to environment (context) changes, has to be able to handle the uncertainties in observations. Moreover, the observations about the health of the system–under–test do not give certainty about an error condition.

Though not addressed in DEMANES, a good example is a hearing aid that adapts is audio processing according its judgement of whether the user is in a train, at home watching TV, listening to speech, biking, etc. Yet it may very well be the case that the system controlling the audio for left ear has a different interpretation of the context than the system handling the right ear. Nonetheless the system must work satisfactorily and certainly avoiding to fall into error modes. Such errors may occur only if a low-bandwidth communication channel between the ears is available that exhibits latency and outages.

In a following subset, we elaborate on our finding for the reliability of presence detection in a building, based on commonly available sensors. Adaptation is again tested via KPIs obtained from the context monitoring, that in short are: *sensor correlation*, *system confidence* and *probability of break occurrence*.

5.3.3.1 Multiple Sensors Correlation

A system with multiple sensors sees correlation between the responses from different sensors. For instance two presence detectors that observe a partly overlapping area see correlated events. Sudden changes in the correlation can be an indicator of a malfunctioning sensor. Hence, in terms of reconfigurability, the correlation between the presence detectors is a measurement that drives the system adaptation according to availability of resources, as follows. If the sensor correlation is low the *self-adaptive system* is able to infer presence using only the non-malfunctioning sensor, while if the correlation is high the same inference is made using all available resources.

5.3.3.2 Implicit and Explicit User Feedback

If the light in a room does not turn on immediately after we press a light switch, our natural reaction is to flip the switch a few more times. Hence unexpected user commands, such as a repeated sequence of inputs or explicit feedback, can be a good indicator of a failing system.

In office rooms with a presence sensor we sometimes face the situation that the lights turn off when we make little motions while working behind a computer or reading a text. Typically the office worker waves his hands to reactivate the system or *awake* the lights. A smart lighting system should monitor whether it regularly occurs, that immediately after the light-on time's out, the sensor sees a user. If so, the system needs to "reparameterize" such waiting time for turning off the lights.

Using this concept of implicit user feedback, intelligent lighting agents for desktop illumination are controlled. The KPI for adaptation in such a case is the system confidence, as follows. Confidence of the system is increased whenever a control action (light turn on, light turn off) is qualified as correct, and conversely, whenever a negative feedback is present to indicate a wrong control action the confidence is lowered. Lights are controlled based on the confidence of the system and negative feedback is interpreted as user movements in front of the sensor node, i.e., as a reaction to a wrong control action. This case in particular will be further detailed on Sect. 5.4.

5.3.3.3 Handling Uncertainties on Sensor Readings

Realtime applications to control appliances, such as intelligent lighting systems, require methods for handling uncertainty at runtime so that user comfort is not altered because of inefficient control actions. An example of light behaviour adaptation in the context of an office, occurs when there are indicators that the next likely event is the occupant leaving the workplace, so that the lights should be turned off. At runtime the complexity of such an adaptation increases because the user comfort is sensitive, so control actions should be taken in a very smoothly reasoning process that is in turns built from several KPIs.

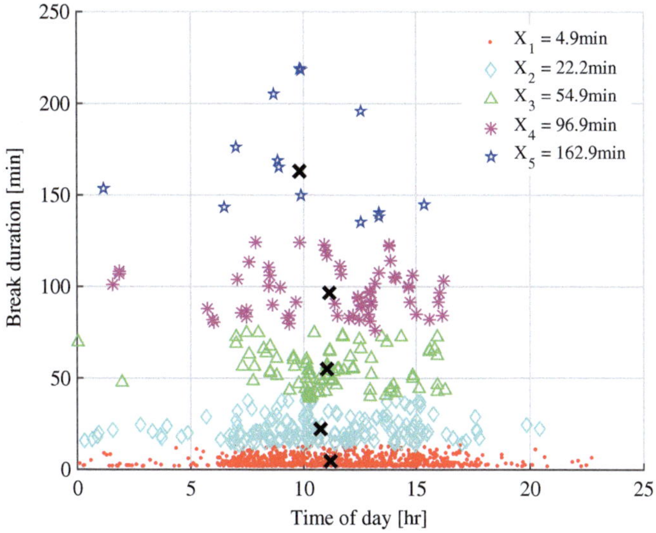

Fig. 5.2 Break clusters identified using Kmeans algorithm on a dataset comprising ≈96 days of continuos presence data recording in an academic working environment

Behaviour of occupants in offices during standard working hours is characterized by periods of work in front of the computer screen, as well as by periods of absence or breaks from the working station. Describing breaks in terms of its duration and probability of occurrence, supports the design of presence-based lighting systems that are able to control lights efficiently. Hence, the break's duration and its probability of occurrence are potential KPIs.

Initially, we analyzed the occupant behaviour in terms of breaks using a dataset of ≈96 days of presence information collected from proximity sensors, in several offices of an academic building [8]. A standard Kmeans clustering algorithm identifies five groups of different meaningful break types n with its respective duration X_n, by means of a centroid analysis. Occurrence probability p_{c_n} is also obtained from the algorithm results (see Fig. 5.2).

Table 5.5 shows that the most concurrent break n_1 is the one with shortest duration $X_1 \approx 5$ min and probability of occurrence $p_{c_1} = 71\%$. Such percentage can be explained if occupant behaviour with respect to office breaks is divided into several categories. For instance, toilet breaks, desk related activities of short duration, phone calls, visits to a colleague's working station at a very close walking distance. Similarly, the group of breaks n_2 with duration $X_2 \approx 20$ min might be associated to a typical occupant behaviour that occurs 14 % of the total recording time, such as coffee breaks.

On the other hand, it is expected that 1 hr absence from the working place, happens with much less frequency depending on the office worker profile. Activities related to this type of office break are for instance, short meetings and lunch break. Finally,

Table 5.5 Probability that a break n of certain duration X_n occurs in the entire dataset using clustering techniques

Break type n	Duration X_n (min)	# of Occurrences	Occurrence probability p_{c_n} (%)
1	4.92	1134	71.14
2	22.22	226	14.18
3	54.94	123	7.72
4	96.89	88	5.52
5	162.94	23	1.44

breaks n_4 and n_5 with duration longer than 1 hr and probability of occurrence less than 6 %, can be interpreted as more complex meetings or activities.

Another potential KPI for the reasoning system is the concentration of the different algorithm centroids between time of the day 10 to 11 hr, indicating that a change in the lighting behaviour might be necessary to couple with the occupant behaviour at that specific time of the day. In this direction Fig. 5.3 shows that the break duration can be modeled as a mixture of overlapping gaussians, in such way that control actions follow smoother transitions according to the break's duration, that can therefore lead to designing *self-adaptive systems* that preserve user comfort.

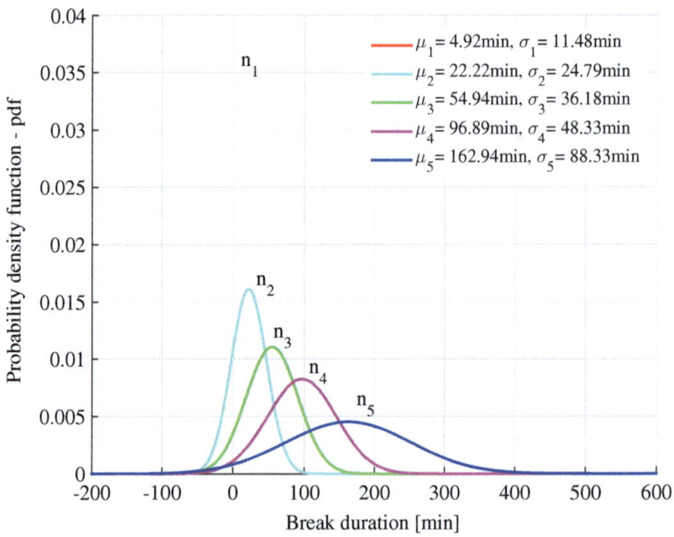

Fig. 5.3 Break duration distribution for each of the considered break types or clusters

5.4 Methods for Testing Under Induced and Normal Operation Conditions

Current available methods for testing software and hardware reliability, include reliability growth models and accelerated life testing. Tools for testing *self-adaptive systems* are still limited, but a combination of available resources could lead to new directions towards the implementation of a V & V framework.

Design and realisation of *self-adaptive systems* is challenging both from an algorithmic as well as from an engineering point of view. Performance analysis with respect to customer requirements is of major importance especially taking into consideration aspects related to dynamic conditions of both the environment and the system itself. Its ability to cope with such conditions and do so within the established user requirements, defines its capabilities of self-adaptation.

Self-adaptive systems can enter fault modes during operation, which can be hard to predict in an isolated testing before installation. Hence during runtime the system must continuously monitor its own health and find ways to detect error positions.

The main methodology for assessing system performance with respect to self-adaptation relates to establishing KPIs that are able to describe system capabilities of adapting (reconfiguring) to various internal and external dynamics. As previously shown in Chap. 1 [10], one can establish metrics that describe the relation between specific resource demands together with their functionalities and the available resources.

Here we accept the use of specific performance indicators or metrics that are derived from measurements in the system and the execution of specific tests, similar to a yearly fire training for employee in a building.

Specifically in DEMANES, devices respond to changing internal and external conditions. If carefully designed, reconfiguration algorithms are usually able to cope with various dynamics by altering system parameters related to for example, scheduling of component execution.

One basic tool employed for storing runtime events is a powerful log and trace module. By annotating various events with timestamps, and the module from which they belong to, we can track the runtime execution of various applications.

One more advanced tool is model checking. Adaptation methods can verify exhaustively and automatically whether this model meets a given specification. If the simulation tools are available, we can verify if specific mathematical models match results obtained via experiments. Later on, we can verify the models for adaptation via deployment analysis.

5.4.1 *Accelerated Life Testing: An Example of a* **Playback** *Feature of the V & V Framework*

Accelerated Life Testing (ALT) or overstress testing is a common practice in which the device is exposed to sufficient overstress to bring the time to failure to an acceptable level. ALT plays the role of *Playback* feature in the V&V framework proposed in Chapter [2]. In this case, the prerecorded or synthetic conditions injected to the testing system are called *stressors*. Some of these accelerated testing conditions or *stressors* may involve a higher level of temperature, pressure, voltage, load, vibration, and so on, than the corresponding levels occurring in normal use conditions. Overstress by simply increasing the loads, for example temperature or electrical power, may drive certain components to new and/or unwanted failure modes that have relevance to the actual field performance.

Some generic rules for performing ALT testing are:

- Find your system stressors or KPIs
  - Field studies or application studies are needed to determine the so-called mission profile or user profile of the product.
  - For example, for SSL products, known stressors are temperature, relative humidity, mechanical forces like vibrations and shocks, electricity and, not to forget, light.
  - By performing Failure Mode and Effect Analysis (FMEA) with experts, the potential stressors of the system can be determined.
- In principle: a component failure $\neq$ a system failure. This particularly holds for SSL products since if one LED is broken it does not mean that the total light output on a system level is insufficient.
- Each component in a system exhibits its own failure behaviour, and needs to be captured by:
  - Experiments by using at least 3 accelerated testing conditions.
  - Numerical/analytical models that describe the reliability physics or physics of failure [11].
- Interactions between the components need to be captured by:
  - Testing sub-systems.
  - Testing the total system.
  - By accelerating environmental user conditions in a physically correct manner.

There are basically two different reliability test approaches: test-to-pass and test-to-fail.

Test-to-pass

Demonstration testing or zero failure acceptance testing is an approach in which a certain number of test cycles is needed without the occurrence of failures. Test to pass only provides pass-fail results; the results do not give any information with respect to the reliability as a function of time (or kilometres or cycles). These limitations are addressed by test to failure.

Test-to-failure

Is an approach in which the tests are continued until at least 65 % of the population failed. This approach will give full information on failure modes, but the limitation could be long duration of the test.

For key-components in any system, it is advised to follow a test-to-fail approach, preferable using meaningful accelerated tests. For the overall system, the test can be divided according to the phases of product development and product released. For the former phase it is advised to follow a test-to-fail approach, while for the latter phase it is advised to follow a test-to-pass approach.

5.4.2 Tools for Testing Runtime Self-adaptive Systems

Context Recognition Network Toolbox, CRNT, is a c++ parameterizable framework that uses *reader* tasks as sensor drivers, so multiple sensors from different manufacturers are read-in at convenient sampling frequencies. The toolbox contains ready-to-use filters and classification algorithms among other components, for signal processing and pattern recognition, so that for instance, the test designers can compose their own testing modules according to the application requirements.

Runtime decision and testing algorithms are deployable in CRNT. An example of presence detection for controlling lighting systems, is illustrated in Fig. 5.4. From this point onwards we will refer to this example to explain the different testing framework components. First let's look at the system functionality.

In this case, the system comprises four main elements: monitoring, classification, decision and actuation. Monitoring tasks are the lower layer and their function is to read all inputs via sensor drivers, that translate environmental and contextual variables into machine readable values. Furthermore, the classification chain incorporates a typical feature extraction and classification process, with filter tasks and binary classifiers, respectively, to select a class, i.e., *presence* or *absence*, from distance measurements that could indicate the presence of objects, persons or none at all. Next, the decision engine selects a control action according to the identified class that takes place at the top layer in the actuation element, i.e., lights turn on or lights turn off.

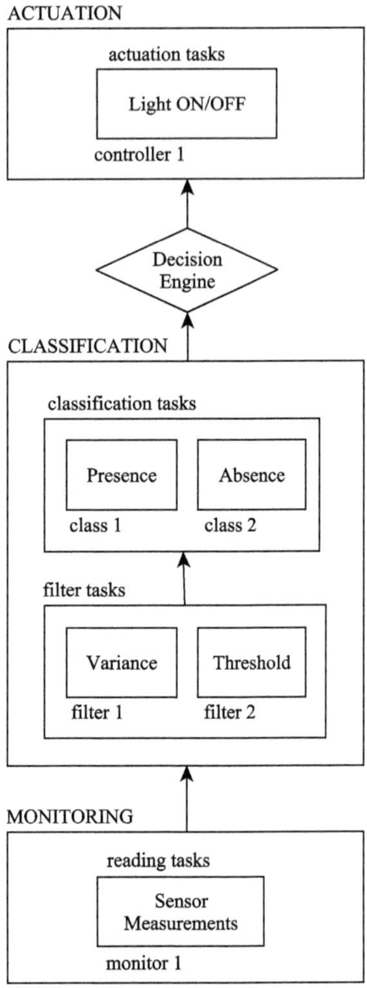

Fig. 5.4 CRN Toolbox runtime decision algorithm for lighting control

CRNT interprets such block scheme using a manually designed Java Script Object Notation (JSON) configuration file, as illustrated in Table 5.6. The configuration file plays a *Test Coordinator* role, giving instructions to the different CRNT modules of when and how to operate, for example, specifying their interconnection. In general, configuration files are divided into two self-explanatory sections, i.e., tasks and connections. Each of the tasks has a set of specifications according to its type, e.g., *Monitor 1* has a reading frequency that differs from that of *Monitor 2*. The connection section is a straightforward translation of the block scheme, in which the only property to specify consist of the source port number that should connect to the correspondent destination port, e.g., an ultrasound detector with an incorporated brightness sensor, acts a *Monitor 1* having two output ports, $port_0$ gives a proximity measurement in distance values, $port_1$ gives a measurement of brightness.

Table 5.6 JSON text interface for a test-coordinator example of a non-adaptive system

```
{
"tasks",
(
{"type": "Monitor1",     "specifications": ...},
{"type": "Filter1",      "specifications": ...},
{"type": "Filter2",      "specifications": ...},
{"type": "Classifier1",  "specifications": ...},
{"type": "Classifier2",  "specifications": ...},
{"type": "Adapter1",     "specifications": ...},
{"type": "Control1",     "specifications": ...}
),
"connections",
(
{"type": "Connection", "source": "Monitor1",        "port": ..., "destination": "Filter1, Filter2",  "port": ...},
{"type": "Connection", "source": "Filter1, Filter2", "port": ..., "destination": "Class1, Class2",    "port": ...},
{"type": "Connection", "source": "Class1, Class2",   "port": ..., "destination": "DecisionEngine",    "port": ...},
{"type": "Connection", "source": "DecisionEngine",   "port": ..., "destination": "Adapter1",          "port": ...},
{"type": "Connection", "source": "Adapter1",         "port": ..., "destination": "Control1",          "port": ...}
)
}
```

The following paragraphs describe the design and implementation of several test cases that incorporate the elements of a standard V & V framework previously described in Chap. 4 [2]. For simplicity, we will refer such description to the example presented in Fig. 5.4.

5.4.2.1 Test Case Formalization

Formalization of the test-case consist of the end-user and test-designer defining a Software Level Agreement (SLA) in terms of Service Level Objectives (SLO), as previously described in Chap. 4 [2]. We need to provide a way for a test designer to define the test cases based on such definition. In brief, that may be e.g. a language or a graphical tool with which the test designer can specify which data, from which nodes need to be collected, what algorithms need to be used for analyzing them, etc.

Yet tools for the automatic translation of the end-user SLAs in terms of SLOs and in turn into system KPIs, are not available. The closest form of creating a bridge between the end-user and the system language are JSON descriptors, often described as an *ideal interchange language*. Many resources are available for syntax checking, design, edition, and so on, of JSON descriptors.[1]

In DEMANES, JSON configuration files are designed according to the type of test. Testing is categorized into two different types according to the action that triggers the V & V framework, thus a test under normal operation conditions is triggered by an adaptation request and similarly, a test under induced conditions is triggered by the *Playback* feature. The test coordinator in form of a JSON file configures the system with an adaptation request to evaluate both, the VZs and the system's

[1] http://www.ethanpublishing.com/.

Table 5.7 Testing case A: System level workflow

Test-case component	Description
FR	A sensor measurement is classified correctly as *user is present* among all measurements (persons, objects, non obstacles) by Classifier 1
NFR	A binary classifier based on variance and threshold filters selects *user is present* class from all sensor measurements (persons, objects, non obstacles)
NFR	The monitored user manually annotates his/her presence instances in the form of time intervals
SLA	Classifier performance is measured in terms of *NFRs* as follows. NFR_1: Classifier precision (P) respect to NFR_3, NFR_2: Classifier recall (R) respect to NFR_3, NFR_3: ground truth (GT) is obtained from user manual annotations of his/her presence instances
SLO	Predefined KPIs are: $KPI_1 = P > 60\,\%$, $KPI_2 = R > 80\,\%$

quality of adaptation and with synthetic/prerecorded data to evaluate the system VZs, respectively (see Chap. 4 [2] for further details on this description).

Following the scheme illustrated in Fig. 5.4 of runtime system the following are examples of test case formalizations for performing testing under normal execution mode, so no *Playback* feature is necessary, according to the testing workflows at system and adaptation levels as specified in Chap. 4 [2]. Functional Requirements (FR) as well as Non Functional Requirements (NFR) are necessary to define SLAs, as detailed in Tables 5.7 and 5.8.

Most presence detection systems use binary classifiers to select from the collected data between the classes: *presence*, *absence*. Hence, critical KPIs determine the classifier performance that can ensure proper operation of further mechanisms, e.g., lighting controllers.

Classifier Precision (P) and Recall (R) are common indicators of such performance, and its calculation typically requires ground truth (GT) measurements, which are collected in many cases from manual annotations of the experimenters, but also via additional monitoring methods such as video recordings in more sophisticated systems.

Self-adaptive systems usually employ feedback mechanisms to dynamically modify adaptation parameters. In adaptive lighting systems controlling the time to turn off the lights or Turn Off Delay (TOD), after a person leaves his/her working station is a significant parameter for energy saving purposes. The shorter the TOD the more energy is saved. At the same time user comfort is affected, thus balancing energy savings and user comfort provides the dynamics for the adaptation. Users can input

Table 5.8 Testing case B: Adaptation level workflow

Test-case component	Description
FR	Adaptive behaviour of the system is consistent and reliable
NFR	Adaptation parameters change less frequently over time and are coherent with feedback input
SLA	Adaptation performance in terms of *NFRs* as follows. NFR_1: Turn Off Delay (TOD) reaches stability after certain period of time $\lim_{t \to +\infty} TOD = 5\,\text{min}$, NRF_2: Confidence reaches stability after certain period of time $\lim_{t \to +\infty} C = 80\,\%$
SLO	Predefined KPIs are: $KPI_1 = 1\,\text{min} < TOD < 5\,\text{min}$, $KPI_2 = 60\,\% < C < 80\,\%$

their feedback via physical light switches or web interfaces when for example, their comfort is altered and as a consequence the TOD of the lighting system is adjusted.

Several KPIs measure the quality of such an adaptation mechanism, in this case consistency and reliability are evaluated by testing the stability of the adaptation parameters, and confidence on the adaptation is measured according to user feedback input.

5.4.2.2 Test Execution

In general, as seen from the examples detailed in Tables 5.7 and 5.8 for testing under normal conditions, the validation and verification framework requires a logging interface to store data, e.g., critical events, monitored data, reports of analysis results, etc.; and a analyzer module in which the testing operations such as viability zone check and adaptation quality check are executed.

Figure 5.5 and Table 5.9 illustrate respectively, the translation of the previously discussed testing cases into a CRNT graphical block scheme and a text interface of a JSON coordinator file.

Runtime reparameterization on CRNT has been previously implemented [8]. Based on the principles proposed in this work, the example of *self-adaptive system* presented in Fig. 5.5 can, as well, be deployed on CRNT. Adaptation takes places whenever it is specified in the test coordinator. In this case, by adjusting timing parameters of the control action demanded via the decision engine.

The additional testing elements, such as feedback and ground truth monitoring can be implemented using the CRNT reading task named *KeyboardReader*. With this feature the monitored user/experimenter can provide a single key or key pattern to provide feedback whenever needed, thus if lights turned off to soon or inefficiently, the system is informed with the key pattern *0 0 0*.

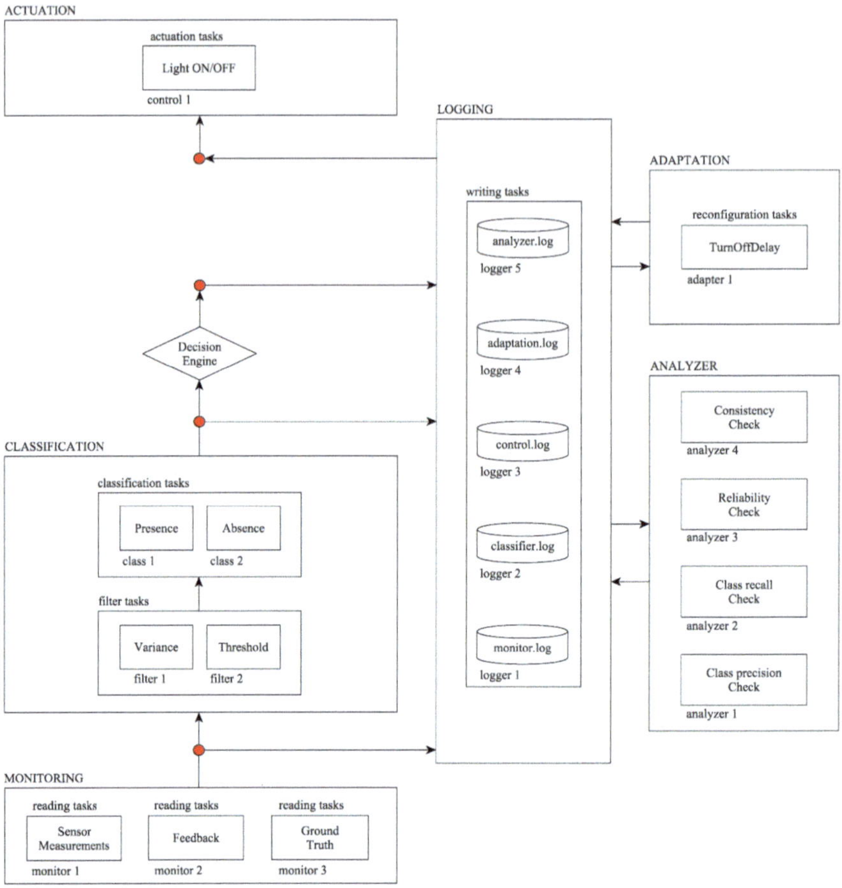

Fig. 5.5 Graphical description of the CRN Toolbox runtime validation and verification framework

Logging is a very common tool in testing, specifically in CRNT the logger module is identified as a *writing* task since it allows to report the different states of the system at each single stage. Logged information is stored in the selected locations or repository destinations for further analysis. Similarly, each component designed within CRNT contains output channels that are used to transmit results, status, and some other information to the subsequent components, that can be interpreted as well as a logging mechanism. Therefore, such output channels are connected to the validation and verification framework via the logging module.

The analyzer checks for the system viability zone and the adaptation quality, using a single analyzer element per each condition to be checked. Usually, analyzers contain relational operations and mathematical modules to execute simple comparisons and calculations. For example, the KPIs precision and recall that measure the classifier

Table 5.9 JSON text interface for a test-coordinator example of a *self-adaptive system*

```
{
"tasks",
(
{"type": "Monitoring",     "specifications": ...},
{"type": "Logger1",        "specifications": ...},
{"type": "Classification", "specifications": ...},
{"type": "Logger2",        "specifications": ...},
{"type": "DecisionEng.",   "specifications": ...},
{"type": "Logger3",        "specifications": ...},
{"type": "Adaptation",     "specifications": ...},
{"type": "Logger4",        "specifications": ...},
{"type": "Actuation",      "specifications": ...},
{"type": "Analizer",       "specifications": ...},
{"type": "Logger5",        "specifications": ...}
),
"connections",
(
{"type": "Connection", "source": "Monitoring",       "port": ..., "destination": "Classification", "port": ...},
{"type": "Connection", "source": "Monitoring",       "port": ..., "destination": "Logger1",        "port": ...},
{"type": "Connection", "source": "Classification",   "port": ..., "destination": "DecisionEng.",   "port": ...},
{"type": "Connection", "source": "Classification",   "port": ..., "destination": "Logger2",        "port": ...},
{"type": "Connection", "source": "DecisionEng.",     "port": ..., "destination": "Adaptation",     "port": ...},
{"type": "Connection", "source": "DecisionEng.",     "port": ..., "destination": "Logger3",        "port": ...},
{"type": "Connection", "source": "Adaptation",       "port": ..., "destination": "Actuation",      "port": ...},
{"type": "Connection", "source": "Adaptation",       "port": ..., "destination": "Logger4",        "port": ...},
{"type": "Connection", "source": "Logger1,2,3,4",    "port": ..., "destination": "Analyzer",       "port": ...},
{"type": "Connection", "source": "Analyzer",         "port": ..., "destination": "Logger5",        "port": ...}
),
}
```

performance, are computed by taking the output of the classification and comparing it with the provided ground truth.

Furthermore, depending on the type of test the V & V framework may require playback modules to inject synthetic or previously monitored data (see [2]), for which case the CRNT contains a reading task named *FileReader* that simulates sensor behaviours by reading log files from previously monitored data or simple synthetic data that is manually input to the system.

5.4.2.3 Addional Testing Tools

Interfaces for testing via screen visualization, at each single stage of the V & V framework are useful in some systems, specially large scale systems, when due to its complexity it is difficult or time consuming to look for specific logged information. Visualization allows a fast checking of the system proper functioning, although it is prompt to the human errors.

For such purpose, CRNT contains the so-called CRNT scope, a graphical testing interface that requires the specification of communication details such as *TCP port* and *IP address* to realize the connection, and visualization details such as the channels or outputs that are desired for the visualization testing and the amount of samples,

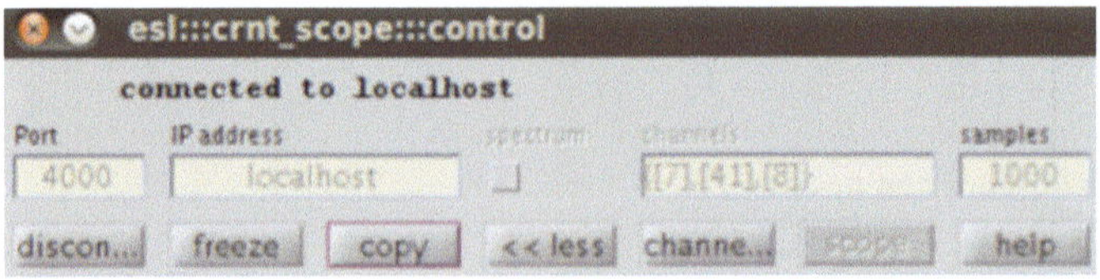

Fig. 5.6 CRN Toolbox scope control interface

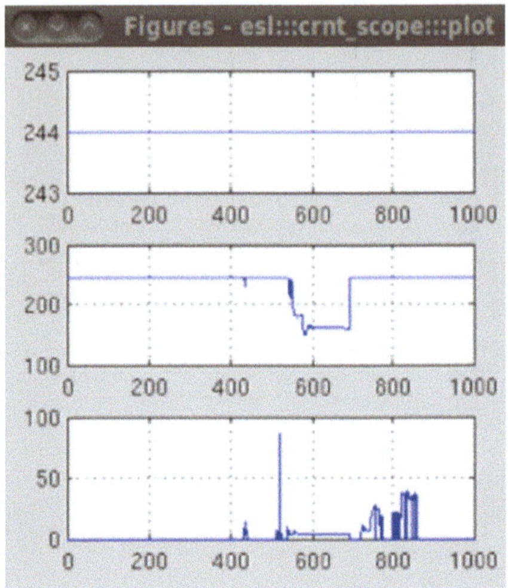

Fig. 5.7 CRN Toolbox scope visualization

as illustrated in Fig. 5.6. TCP writer and reader tasks for interconnection to external programs like the CRNT scope are already available in the toolbox contents.

Figure 5.7 shows an example of a monitoring sensor node, that in this case displays its address number *244* (at the top), and monitors two variables: brightness (in the middle) and distance (at the bottom). Whenever activity is reported in the distance channel, in this case showing that a person is interrupting the field of view of the sensor node, the brightness channel reacts decreasing its value as a measure of the shadow produced by the blocking person or object.

Large scale distributed systems take advantage of TCP writers and readers as they allow remote testing, simply by specifying the source or destination IP addresses to stream data back and forward. This interconnection to the outside world is frequently used for executing more complex adaptation processes that are not possible at the node level, by interconnecting one or several systems stages to external powerful programs for signal processing such as Matlab.

Other frequently used tools of such a type of systems are the channel selectors and the channel mergers. The former one represents a distribution tool, in which a module output can be connected to several other components. For example, a monitoring node with three output channels named (1) address, (2) distance and (3) brightness, as the one showed in Fig. 5.6, can be connected via its channel (2) to a filter module and to a logger module, while it is connected via its channel (1) to a diagnosis module that checks the sensor identification. The latter feature greatly supports distributed systems by merging sensor node channels and tunnelling them to the other modules or nodes or to a repository via logger components. It mimics the function of a synchronizer. For example, in case of a lighting control system that bases its control decision on the presence information provided from three nearby monitoring nodes of a localized desk lamp, the merger function output will look like *t 1 0 1*, meaning that the monitoring nodes at a time *t* reported *presence*, *absence* and *presence*, respectively.

In case of distributed systems where the nodes are capable of simple signal processing tasks and decision making, the presence information can be reported back or shared with such nodes so that they individually decide between *presence* or *absence* classes and subsequently produce a control action for the localized desk lamp.

5.5 Conclusions

Testing is an essential part of any complex systems that combines hardware with software. The test effort becomes harder with an increase of the system complexity and with the number of variations that can be out in the field. Particularly if distributed systems reconfigure themselves in runtime, this poses challenges. For example, in identifying the proper KPIs and collecting the respective variables that carry the information which will allow the VZ check and the adaptation quality check.

In the DEMANES research and writing of this chapter, we experienced that the insights in how software testing of runtime reconfigurable systems is still limited. Several components need to be interconnected to achieve reasonable testing results: end-user descriptions, JSON translators and CRNT interpreters in form of system monitors and analyzers capable of adaptation, also in distributed forms.

This chapter gathered insights for a number of relevant testing aspects. A careful design of the data gathering from tests is essential, and therefore a careful design of logging strategies is fundamental.

References

1. X.J. Fan, W.D. van Driel, *Solid State Lighting Reliability: Components to Systems* (Springer, New York, 2012)
2. R. Kotian, Design Process Flow: Methodology and Guidelines
3. M. Lyu, *Handbook of Software Reliability Engineering* (McGraw-Hill and IEEE Computer Society, New York and Los Alamitos, 1996)
4. B. Randell, C. Landwehr, A. Avizienis, J.-C. Laprie, I.E.E.E. Trans, Dependable Secure Comput. **1**, 11 (2004)

5. H. Giese, H.A. Müller, M. Shaw, R. De Lemos, *Software Engineering for Self-Adaptive Systems II* (Springer, Berlin, 2013)
6. W. Nelson, *Accelerated Testing Statistical Models, Test Plans, and Data Analyses* (Wiley, New York, 2004)
7. D. Bannach, O. Amft, P. Lukowicz, IEEE Pervasive Comput. **7**(2), 22 (2008)
8. P. Jaramillo Garcia, L. Lopera Gonzalez, O. Amft, J. Ambient Intell. Humanized Comput. **6**(2), 207–221 (2015). doi:10.1007/s12652-014-0222-2
9. H. Pham, *Software Reliability* (U.S. Government Printing Office, 2000). http://books.google.nl/books?id=TI0Sj6er8UEC
10. Z. Papp, Model-Based Engineering of Networked Embedded Systems (NESs)
11. Calce: Center for Advanced Life Cycle Engineering. http://www.calce.umd.edu/

Chapter 6
An Illustrative Application Example: Cargo State Monitoring

Coen van Leeuwen, Vicente Hernández Díaz, Roshan Kotian, Raúl del Toro Matamoros, Zoltan Papp and Yolanda Rieter-Barrell

Abstract The previous chapters in this book provide the foundations and a brief description of the DEMANES tool chain. This chapter describes a real self-adaptive system developed using the DEMANES tool chain. This chapter focuses on the design and implementation stages of a real use case development. The use case under study is a subsystem, called Cargo Monitoring System (CMS), that monitors the state of the container cargo and pushes all the data to a back office infrastructure for further processing. The containers can be on a truck, a train, or any other appropriate transportation means, or stacked in a container terminal or a cargo ship. A WSN will measure physical magnitudes (temperature, humidity and so on) inside a container and will forward data to others processing nodes in the CMS network. The CMS has to meet several self-adaptive requirements. For instance, the parameters of the CMS elements that monitor the container cargo state are reconfigured accordingly to adapt to internal or external changes (e.g. a low battery level or a container temperature out of the adequate bounds), and the CMS adapts the WSN nodes power transmission to save energy while providing an acceptable quality of service.

C. van Leeuwen · Z. Papp · Y. Rieter-Barrell (✉)
TNO, Oude Waalsdorperweg 63, The Hague, The Netherlands
e-mail: yolanda.rieter@tno.nl

C. van Leeuwen
e-mail: coen.vanleeuwen@tno.nl

V.H. Díaz
UPM, Calle Alan Turing 3, Madrid, Spain
e-mail: vicente.hernandez@upm.es

R. Kotian
TU/e, Den Dolech 2, Eindhoven, The Netherlands
e-mail: r.kotian@tue.nl

R. del Toro Matamoros
UPM, Ctra. de Campo Real km. 0.200 La Poveda, Arganda del Rey, Madrid, Spain
e-mail: raul.deltoro@car.upm-csic.es

Z. Papp and G. Exarchakos (eds.), *Runtime Reconfiguration in Networked Embedded Systems*, Internet of Things, DOI 10.1007/978-981-10-0715-6_6

6.1 Problem Definition

The example application presented in this section proposes a CMS used to monitor the state of container cargo at real-time during transportation from its original to its final destination.

A large amount of goods is transported across the world in sea containers. An efficient and effective logistics chain is of importance to ensure goods reach their final destination in the correct condition. During transportation the container and the goods therein can encounter expected and unexpected events.

A container will change modality several times: part of the transport may be by ship and other parts could be by truck, train or barge. The environmental conditions will change if, e.g. a container starts it's trip in Brasil and ends it in Europe. These are expected events and can be planned for up front. Unexpected events have a larger impact as they can cause goods to be damaged or perished. Examples of such events are transportation delays, severe weather conditions and accidents causing excessive temperature variations and vibrations in the container or simply a late delivery. When fresh goods such as vegetables, meat and fish are transported, an efficient logistics chain is of importance to ensure the goods do not perish and reach their final destination in the exact condition required by the buyer. But also other types of cargo, such as electronics, must reach their final destination in one piece. Cargo transportation of goods such as dangerous fluids, medicine or valuables sets higher demands on the logistics chain and especially security.

The current logistics planning processes at terminals and transport companies are highly inflexible. Container transportation is planned on a daily basis and one to several days up front. From that point onward the daily planning remains static and it is very hard to react to unexpected events during the container trip. Human intervention may occur and is based on personal experience and ad hoc activities: a truck driver may decide to change the order of work items for that day if a traffic jam occurs, or a truck driver may call his planner to send another truck if his breaks down. Real-time actionable information based on the actual and expected state of the cargo, is not readily available to the human actors in the logistics chain preventing them from reacting to the real-time situation effectively and efficiently. This inflexibility in container transportation planning results in goods arriving at their final destination behind schedule and possibly also perished or damaged goods.

Current IT technology allows for quite detailed tracking and tracing of containers and the goods they are carrying. The CMS presented in this chapter consists of a set of sensor nodes placed in a container. The sensor nodes monitor the cargo state continuously during the container trip from beginning to end. In the case of a relevant event, the alert is communicated outside the container to the relevant user in real-time. The user can be one of several stakeholders in the logistics chain: the owner of the goods, the shipping company or the transport company. Such real-time information can be used to adapt the logistics planning process by e.g. prioritising the container handling at the terminal and prioritising the container pick-up by the transport company to maintain the goods.

The CMS is deployed in the container during loading at it's starting point. Either dedicated containers are equipped with such a system or the system has a plug and play nature, allowing for a simple deployment and automatic set-up by an inexperienced user. The system should be applicable to a range of container sizes and types of cargo. In other words, the system should be deployable in a 20' container and a 40' container. The system should be capable of monitoring the state of bananas but also LED-screens. Once the door of the container closes, all users should be able to receive cargo state alerts remotely.

During the trip the container will change transportation means several times. The container is transported to and from the hinterland by means of a truck, barge or train. From port to port the container will be stacked on a ship in between other containers, as is also the case on a terminal. In all cases the cargo monitoring system must be able to deliver real-time information on the state of the cargo. More importantly, the system must remain functional for the entire container trip, keeping in mind that the length of the trip can vary in length between days and months. Finally, the cargo state information must be actionable. This means the cargo state must be updated regularly and should be accurate to capture all relevant alerts and result in a reliable alert system.

The following user requirements can be identified for the cargo monitoring system.

- Cargo state monitoring: the system must determine the state of the cargo for any type of cargo or container.
- Self-configuration: the system must configure itself autonomously during deployment and at runtime.
- Seamless communication: the system must communicate real-time cargo state alerts to the cloud from all transportation means, whether the container is stationary or moving, in a stack or stand alone.
- Container trip lifetime: the system must remain operational for the entire container journey.
- Remote cargo state distribution: cargo state alerts must be received by users remotely at real-time.

6.2 Design Challenges

A number of system requirements can be derived from the user requirements described in the previous section and these requirements set design challenges for the CMS.

The required CMS asks for an algorithm which calculates the cargo state using sensor readings from the sensor nodes in the container. Once the cargo state has been determined, the system must determine whether the present state asks for an alert or not. As the system should be able to do this for different types of cargo, the cargo state monitoring must be configurable for different cargo models. The user should only have to indicate the type of cargo loaded into the container and the system will configure itself accordingly. The new cargo model could also be accompanied by other cargo specific settings, such as update rate and alert threshold. In other words, banana's ask for a completely different alerting system than LED screens.

Self-configuration of the system asks for a plug and play functionality of the system at deployment and runtime. This can be accomplished with autonomous detection, adding and removing of nodes and services in the cargo monitoring system. The nodes should autonomously detect each other, communicate and distribute tasks independent of their exact location. Depending on the location of a truck or container the available nodes can change, e.g. on a truck, ship or train other nodes may be available to include in the network at the container terminal, in a container stack other cargo monitoring systems may be available. Additionally, nodes may malfunction. Not only should the system be able to detect this, the system should be able to redistribute the tasks of that node to other nodes and re-route communication while at the same time maintaining it's primary functionality: cargo state estimation. This flexibility will make deployment of the system easier but also allow the system to be used for several container sizes, adapt to unexpected events and malfunctions.

Seamless communication means the system should be able to communicate the cargo state alerts from the inside of the container to the outside world at all times, independent of it's location. Depending on where the container is the communication channels and routes in/or outside the container will vary. Reconfiguration mechanisms can be applied to allow the system to adapt to the optimal or at least available communication routing. On a ship the communication routing will be different than on a truck. A container stacked at the terminal may be able to use a dedicated gateway at the terminal, which is not available on a ship or truck. The system will always strive for seamless communication driven by availability resulting in a robust uplink to the exterior of the container.

An optimal system lifetime, i.e. as long as the container trip lasts, introduces the need for some kind of power management. The battery operated nodes in the system should remain operational during the trip. The battery lifetime of the nodes in the system is influenced by the footprint of tasks running on the nodes and the task execution frequency. Specifically, the following properties can be adapted to influence the battery lifetime: sampling frequency of the sensing/monitoring tasks, choice of algorithm for data fusion, transmission power or transmission frequency of the nodes. There is a trade-off to be made between saving the battery lifetime of the nodes or some of the nodes and still being able to delivering reliable, relevant cargo state alerts. The adaptable properties listed above influence both the battery lifetime as the reliability and relevance of the cargo state alerts. runtime reconfiguration mechanisms include monitoring, reasoning and control. The monitoring could be a battery monitor or an uplink monitor, the reasoner could perform the trade-off between battery level, transmission power and state estimation performance and finally, the control could be a transmission power control.

6.3 System Design

In this section we will present the physical, task and mapping model of our CMS design following the design methodology described in part I. The figures are generated using the Metaedit+ tooling.

6.3.1 Task Model

The task model presents the tasks that have to be run in the system to accomplish the desired response. The tasks in the task model could be described in a more detailed way using behavioral models. The task model reveals a coarse-grained point of view of the system, readily understandable at a glance, and the set of behavioral models that extends a task model provides a fine-grained view.

Figure 6.1 depicts the task model of the CMS. The tasks in the that figure could be grouped in reconfiguration tasks and primary tasks:

- Reconfiguration tasks: those that carry out the reconfiguration of the system to achieve the self-adaptation functionalities. The foundations are fully described in [2].
  - The goal of the tasks group called *stateReconfGroup* is to reconfigure different elements in the system to estimate accurately the state of the container (based on the measures samples) at any point in the container, independently of the changes in the elements in the system (crashing nodes, interferences, failing links and so forth).
  - The tasks group called *powerTransmissionReconfGroup* adapts the power transmission of the nodes in a WSAN, to achieve the best trade-off between the energy consumption and the network connectivity, considered as the number of packets the WSAN sink receives, overcoming changes in the system.
  - The tasks groups called *nodeStatusMonitoring*, *CMSStateFusion*, *CMSStateInterpreter* and *CMSStateEventHandler*, act on different elements in the CMS to recover from unexpected changes in the system (see Table 6.1 for a detailed description of each). More precisely, they control some parameters of the nodes: the sampling rate, the power transmission, -to ensure a specific RSSI in the receiving nodes-, the link quality with the gateway and the storage performance. To accomplish this, the battery level, the RSSI of the links with other nodes, the quality of the link with the gateway and the instant storage capabilities have to be monitored. In this context, the sampling rate refers to that for monitoring, not for measuring physical magnitudes.
- Primary tasks: those that are not directly involved in the self-adaptation accomplishment of the system but are tied to the specific behaviour expected from the CMS application:

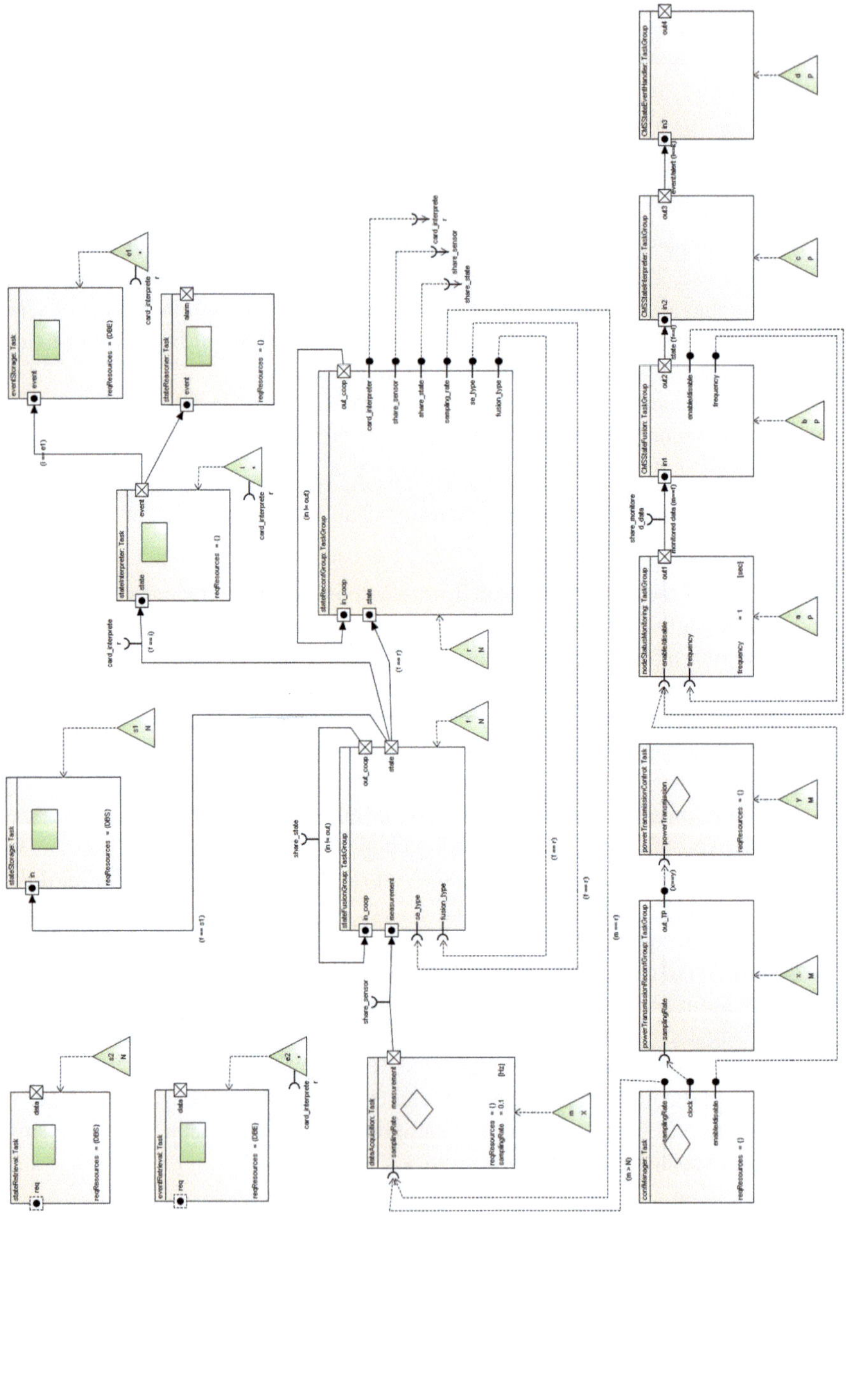

Fig. 6.1 CMS task model

Table 6.1 Tasks groups for preserving links quality, storage and battery level

Task group name	Description
nodeStatusMonitoring	Monitoring of the battery level in each node, the links quality between nodes pairs, the instant storage capability and the link quality with the gateway
CMSStateFusion	Estimation of the state of the CMS based on the data monitored by the *nodeStatusMonitoring*. This task group controls the sampling rate for the data monitored by *nodeStatusMonitoring*, enables or disables the monitoring tasks that make it up and stores the CMS state
CMSStateInterpreter	Analysis of the CMS state for determining when it is unsafe
CMSStateEventHandler	Accomplishing actions to reconfigure CMS elements whenever *CMSStateInterpreter* triggers an event or alarm

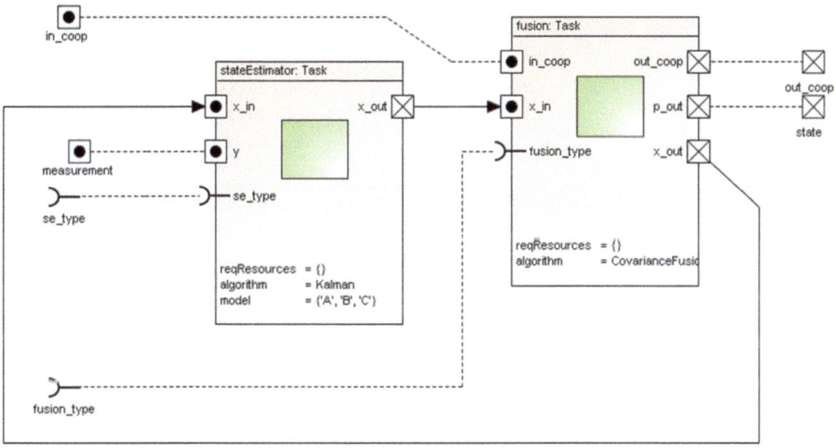

Fig. 6.2 Tasks that make up the *stateFusionGroup* taskgroup

The rest of this subsection details the tasks that make up each task group in the Fig. 6.1.

In Fig. 6.2 the *stateFusionGroup* is depicted. The function of the complete task group is to process the local measurements taken in the *dataAcquisition* tasks into an estimated state of the complete container. In order to estimate this complete state, there are two tasks in the group, running signal processing algorithms.

The *stateEstimator* task receives from the *dataAcquisition* task the raw measurements and based on the previous state and a model of the container, estimates an updated state of the container. It can do this by using different state estimation algorithms such as a Kalman filter or a Particle filter. Which algorithm is eventually used is not determined at design time, but is configurable by the reconfiguration tasks (Table 6.2).

Table 6.2 Primary tasks of the CMS

Type	Task	Description
Monitoring the state of the containers and the trucks	*dataAcquisition*	Data acquisition from the sensors in the WSAN. The sampling rate for some WSAN nodes will be reconfigured at runtime and for others will be configured by the *confManager* task
	stateFusionGroup	Estimation of the state of any point inside the container based on the measures from some sensors deployed in it. The estimation algorithm could be reconfigured
Managing alarms	*stateInterpreter*	Filtering out of the collected data to detect and warn about remarkable situations
	stateReasoner	Making decision task for triggering alarms based on the results from the task *stateInterpreter*
Registering and retrieving data	*stateStorage*	Registration of the containers and trucks state
	stateRetrieval	Retrieval of the containers and trucks state
	eventStorage	Registration of the events related to the containers and trucks state
	eventRetrieval	Retrieval of the events related to the containers and trucks state
Configuring some elements of the system	*confManager*	Configuration of some elements of the system
Actuating on elements properties	*powerTransmissionControl*	Adjust of the power transmission of nodes in the WSAN based on the output from the task *powerTransmissionReconfGroup*

The *stateFusion* task receives the estimated state from the *stateEstimator* and from other *stateFusion* tasks in the system. It combines all estimated states into one estimation of the complete container. In the same way as for the state estimation task, there are multiple algorithms that could implement this task, and only during runtime it is decided which algorithm is used. For instance there is the Consensus (or Averaging) method, but also more complex methods such as the Covariance Intersection or the Ellipsoidal Intersection.

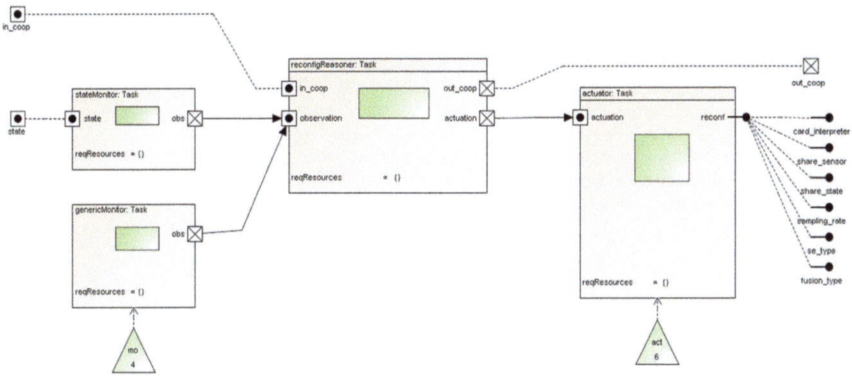

Fig. 6.3 Tasks that make up the *stateReconfGroup* task group

Figure 6.3 details the *stateReconfGroup* task group. The function of this task group is to make sure the tasks performing the state estimation run in such a way that they perform well according to the system specifications. The setup of this task group is similar to the MAPE-K approach [1]. It defines tasks that monitor the system, those that analyze and plan the reconfiguration, and those that execute the reconfiguration actions.

The monitoring tasks are the eyes of the state reconfiguration group, and observe the internal status of the system. The *stateMonitor* receives the state as estimated by the *stateFusionGroup*, and analyses will pass it on to the next task. Other monitors (not specifically drawn in Fig. 6.3) could monitor the battery level of a node, the communication bandwidth, or other measures of the systems performance. All the acquired information is passed on to the reconfiguration Reasoner.

The *reconfigReasoner* task is the brain of the state reconfiguration group. It is given the responsibility to analyze all information gathered by the monitors and come up with actions (or a plan) that will make sure the system requirements are met. One implementation of such a reasoner could be a rule base which will trigger certain actions based on logic rules as described in [2]. Another approach is to run a discrete optimization algorithm which would do design space exploration in order to find a (near) optimal set of parameters according to a designer-specified performance function.

Finally the actuator tasks act as the hands of the group, and have the know-how of how to change the system components so that they correspond to the configuration that the reasoner has come up with. In Fig. 6.3 they are not named, but one of the actuator tasks could be specifically designed to start and stop the state interpretation tasks on other nodes, and thereby redefine in runtime the cardinality of the state interpretation task. A simpler actuator would be one that simply changes the frequency parameter of a *dataAcquisition* task. The exact behaviour of these actuator tasks are very dependent on the implementation of the runtime services (middleware) in the system.

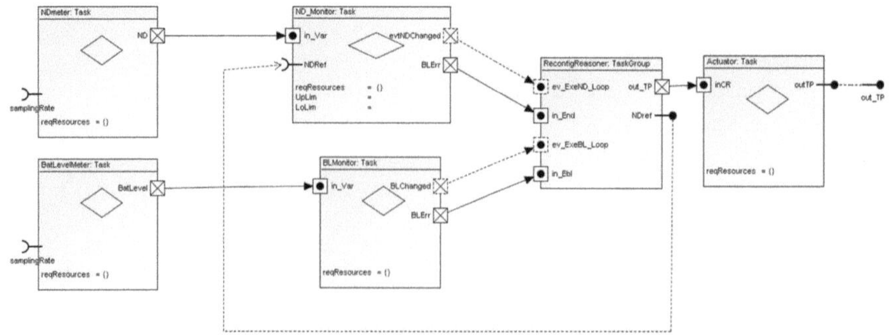

Fig. 6.4 Tasks that make up the *powerTransmissionReconfGroup* task group

Figure 6.4 details the *powerTransmissionReconfGroup* task group. As above mentioned, it depicts a task model for implementing a transmission power control in order to improve the overall connectivity of the network. The algorithm could be deployed in every node of the network and it will reconfigure the transmission power at runtime. The self-adaptive algorithm, which is based on a feedback control system design method, aims to control the transmission power of each node to manage the number of neighbors of each node (node degree from now on), in order to recover the link when its neighbors fail. On the other hand, it is more likely that the node consumes more energy if the node degree is high. Therefore, in order to avoid the node degree becoming too high when the node energy becomes low, the algorithm also controls node degree to balance the connectivity and the energy consumption.

In others words, the basic idea of the control system is that if the node degree is higher than the desired node degree, then the transmission power is decreased; if the node degree is lower than the desired node degree, the transmission power is increased. The desired node degree is associated with the energy of the node. The controller (i.e., fuzzy logic based controller) decides how fast and how long the transmission power changes.

The model consists of the following tasks:

- **Meter tasks**: this type of task should access the node physical resources. The model has two tasks of this type, to measure the node degree in the node (this information should be already stored by the routing protocol) and the actual battery level, respectively.
- **Monitoring tasks**: this type of task compares if the difference (i.e., error) between the measured variable (output from a meter task and input of the monitoring task) and the reference value (it is a reconfigurable input variable of the task) for this variable is inside a specific region (it is bounded by task parameters UpLim and LoLim). The task calculates the error and if the error is out of bound, it generates an event to execute the reasoning task. The model has two monitoring tasks, one for the node degree and another one for the battery level.
- **Reasoning tasks**: this category of task will execute the decision-making algorithm of the reasoner to calculate the appropriate variation of a variable, according to the error value. Specifically, there are two reasoning tasks, one for each measured

variable. The decision-making algorithm is based on fuzzy logic, and this requires that each task has the knowledge base (e.g., rules, internal functions parameters, etc.) as a property. Reference values (node degree reference and battery level reference) are also reconfigurable properties of the task and outputs of the task. The task for reasoning about the node degree error also has the variation of transmission power as output. The basic idea is to vary the transmission power around a specific nominal value adjusted at design time.

- **Actuator task**: this task has the variation value of power transmission as input, and it calculates the absolute value of this parameter and tries to keep it inside its bound as well.

The complete task group has an input reconfigurable parameter. It is the sample rate, which determines its execution frequency.

Figure 6.5 shows in detail the tasks group *nodeStatusMonitoring*. The tasks *node2nodeLQ*, *batteryMonitor*, *storageMonitor* and *uplinkMonitor* are run in each node and they monitor respectively the quality of the nodes link with others neighbors, the battery level, the available storage capability and the quality of the link with the gateway.

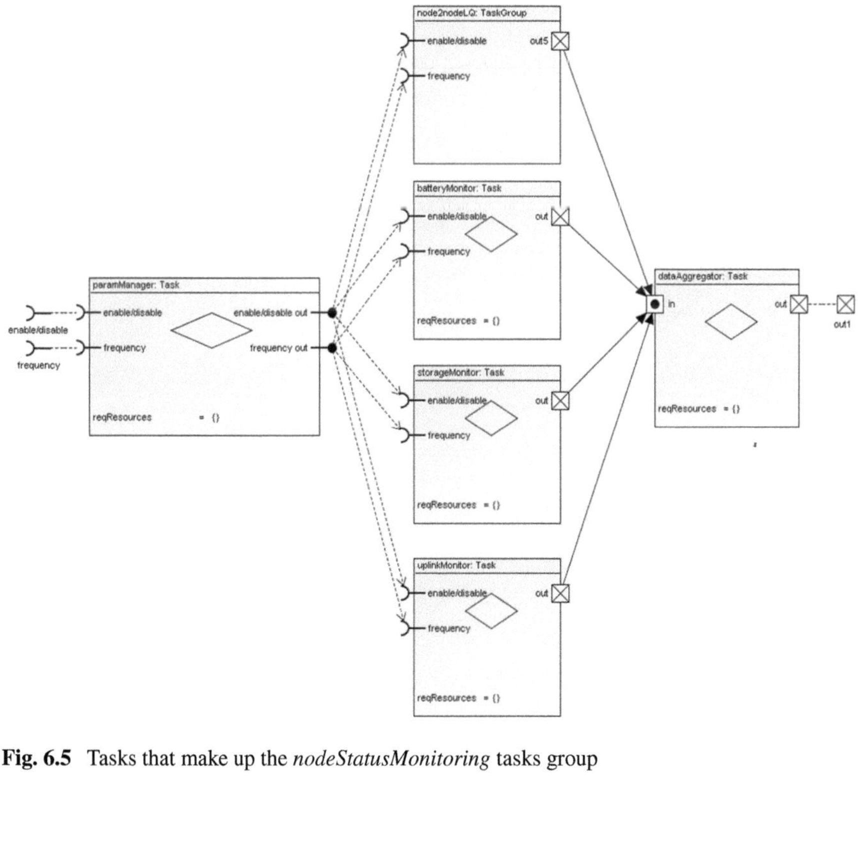

Fig. 6.5 Tasks that make up the *nodeStatusMonitoring* tasks group

The monitoring sampling rate depends on the parameter ***frequency***. On the other hand, any of the mentioned tasks can be enabled or disabled by either the user or the tasks group *CMSStateFusion*, by means of the parameter ***enable/disable***.

The task *dataAggregator* gathers the monitored data and provides a suitable summary for being processed by the tasks group *CMSStateFusion* that is depicted in Fig. 6.6.

The task *filter* filters out the monitored data. The filtering rules are reconfigured at runtime, depending on the CMS state, and are managed by the task *CMSStateFusionReconf* by means of the ***filtering_rules*** parameter. The task *CMSStateBuilder* calculates and estimates the state of the CMS. That state will be stored by the task *CMSStateStorage*, will drive task *CMSStateFusionReconf* reconfiguration activities and will be an output for other tasks groups in the system, e.g. *CMSStateInterpreter*. The task *CMSStateFusionReconf* manages the parameters ***enable/disable***,***frequency*** and ***filtering_rules*** that let it, respectively, enable or disable the monitoring tasksindextask!monitoring accordingly, act on the monitoring sampling rate and decide the filtering rules.

Figure 6.7 describes the tasks group *CMSStateInterpreter*. The *CMSStateInterpreter* analyses the instant node state provided by the *CMSStateFusion* task group and the previous ones, provided by the task *CMSStateRetrieval*, and triggers events and alarms appropriately, which will lead the actions carried out by the tasks group *CMSStateEventHandler*, illustrated in Fig. 6.8, to correct the node state.

The *eventReasoner* task determines which actions have to be accomplished depending on the events and alarms generated by the *CMSStateInterpreter* task. The actions will be actually carried out by the *CMSStateActuator* task.

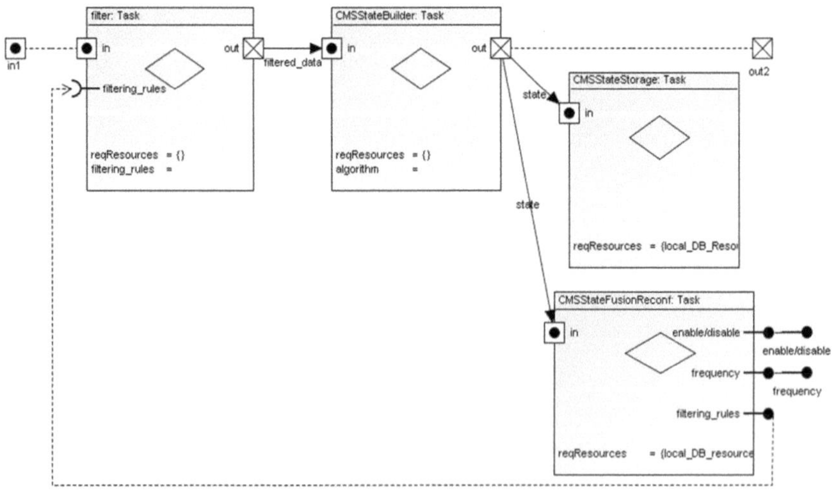

Fig. 6.6 Tasks that make up the *CMSStateFusion* tasks group

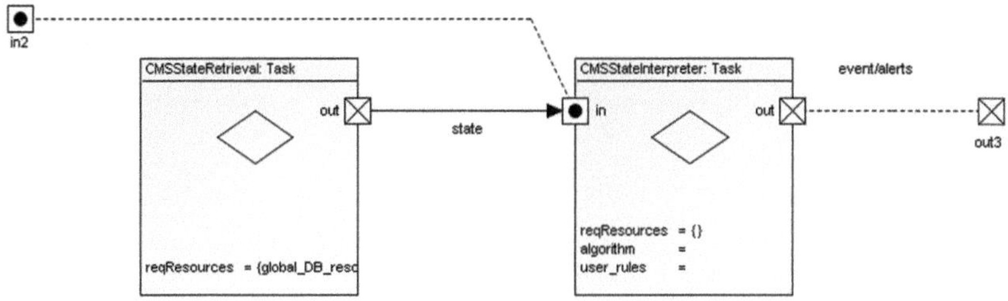

Fig. 6.7 Tasks that make up the *CMSStateInterpreter* tasks group

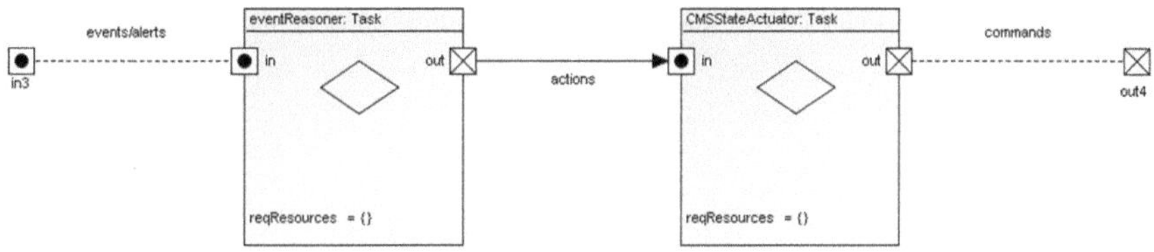

Fig. 6.8 Tasks that make up the *CMSStateEventHandler* tasks group

6.3.2 Behavioral Model

A task behavioral model provides a more detailed view of a task. This section will go deeper in the tasks in Fig. 6.4 and will provide the behavioral model of those closely related to the system reconfiguration capabilities. The goal of those tasks is described in [2], and depicted in the Fig. 6.9. The transmission power (*CR*) will be reconfigured taking into account each node number of neighbors (*ND*) and battery level (*E*), in order to guarantee the WSAN connectivity.

The tasks *NDmeter* and *BatLevelMeter* in Fig. 6.4 measure respectively the number of neighbors and the battery level at each node. Their behavioral models are quite similar and only the one for the *NDmeter* is shown in Fig. 6.10.

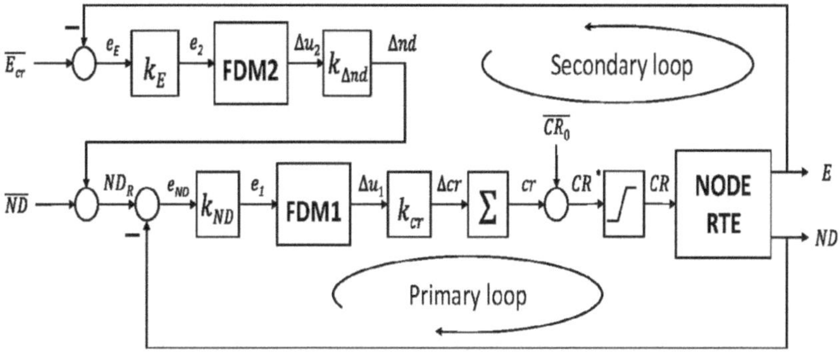

Fig. 6.9 Transmission power reconfiguration

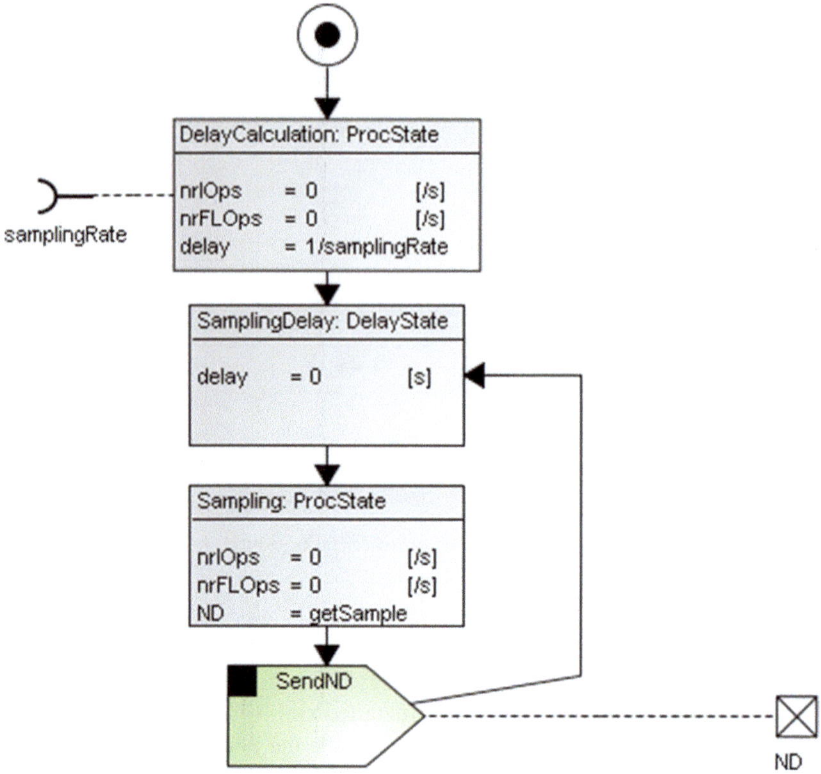

Fig. 6.10 Behavioral model for the task *NDmeter*

The task stops for (1/*samplingRate*) seconds before measuring the number of neighbors, measures it, sends the result to other tasks through the port *ND* and loops again. The sampling rate is a reconfiguration parameter.

The tasks *NDMonitor* and *BLMonitor* have to calculate the error between the measurements from the previous tasks and the corresponding reference values (e_E and e_{ND} in Fig. 6.9), and to trigger an event whenever the error absolute value goes out of specific bounds. The reasoner will capture the event and will make a proper decision to reconfigure the system and correct the deviation. Both tasks are quite similar and the behavioral model of just one of them is provided in Fig. 6.11.

The tasks group called *ReconfigReasoner* in Fig. 6.4 is made of two tasks, as depicted in Fig. 6.12, *NDReasoning* y *BLReasoning*. They implement respectively the *Primary loop* and the *Secondary loop* in Fig. 6.4. The first one will reconfigure at runtime the power transmission of the node whenever the number of neighbors deviates from the *NDRef* reconfiguration parameter. The second one will adjust *NDRef* whenever the node battery level drops below a specific reference value, estimated at design time.

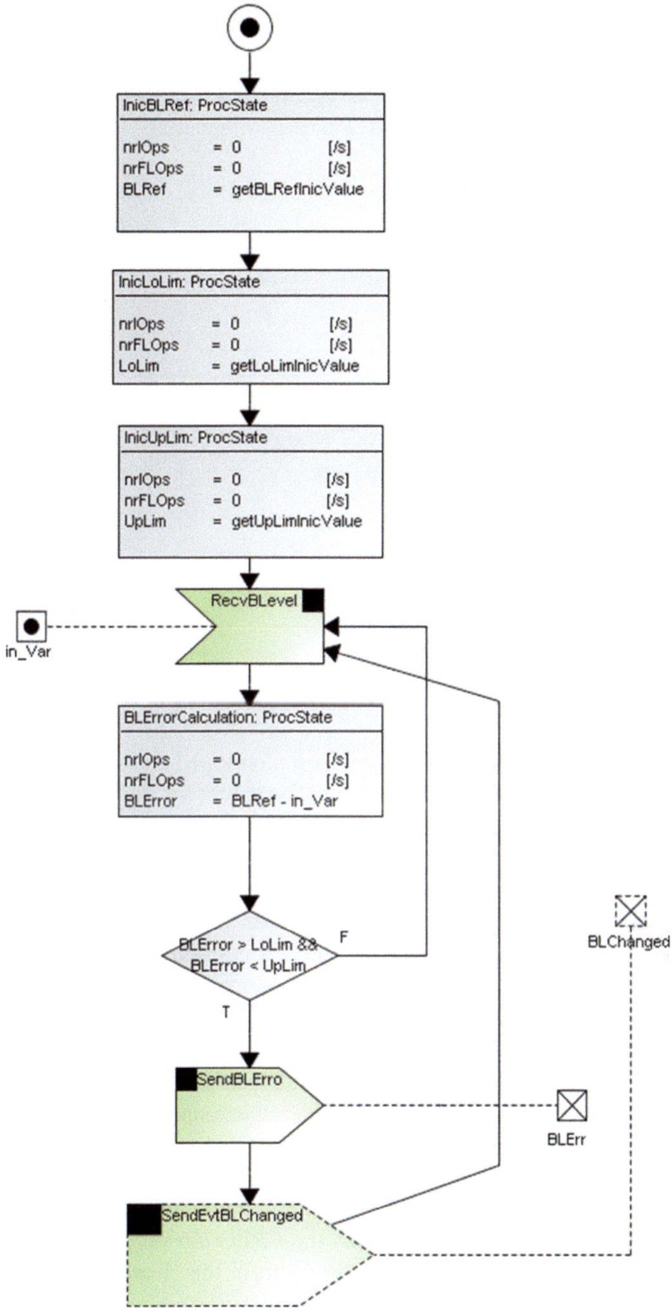

Fig. 6.11 Behavioral model for the *NDMonitor* task

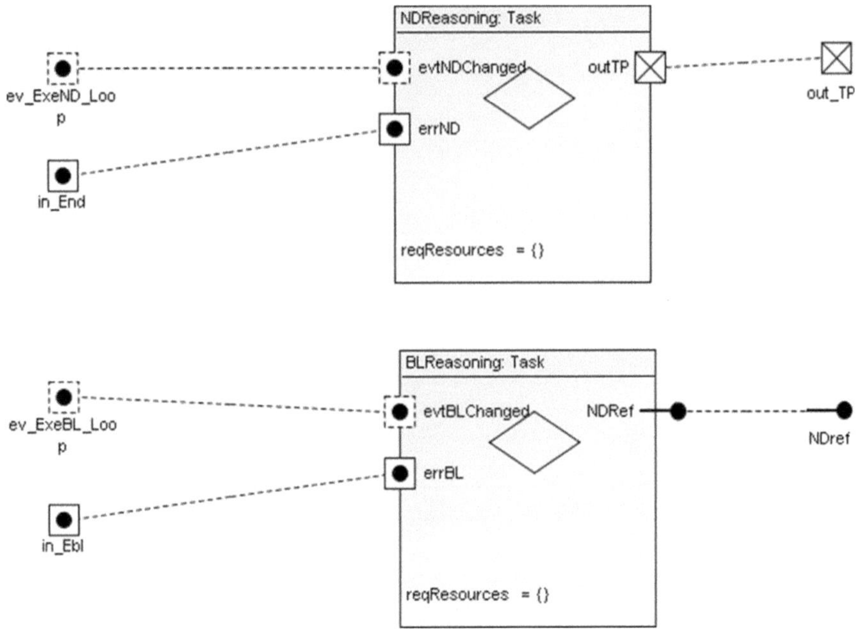

Fig. 6.12 Tasks that make up the *stateFusionGroup* task group

The behavioral model for the *Primary loop* mentioned above, i.e. the one for the *NDReasoning* task, is described in Fig. 6.13. The values for Fig. 6.9 parameters k_{ND}, k_{cr}, $(\overline{CR_0})$, CR_{MAX} and CR_{MIN}[1], are first initialised. The task will then wait for an incoming event from *ND_Monitor* task indicating that the number of neighbors has changed significantly and it makes sense to check if reconfiguration is needed. The difference between the current number of neighbors and the desired one (errND) will be received from the same task. Such error will be amplified (depending on the desired dynamics for the reconfiguration) and will be the input to the fuzzy reasoner FDM1. FDM1 will provide how much the power transmission should be increased to correct the error. The increment to be applied will be amplified (depending also on how the deviation should be corrected) and added to the last increment. So that *Summation* processing state output provides the global increment since the task beginning. The *Summation* output is added to the initial node power transmission and the result is the new power transmission. As a node power transmission does not range from $-\infty$ to ∞, a limiter processing state is needed to guarantee that node CR_{MAX} and CR_{MIN} bounds are never exceeded. Finally the *CRLimiter* processing state output is sent to the *Actuator* task that will actually perform the change in the node power transmission.

[1] CR_{MAX} and CR_{MIN} refer respectively to the maximum and minimum power transmission for a node.

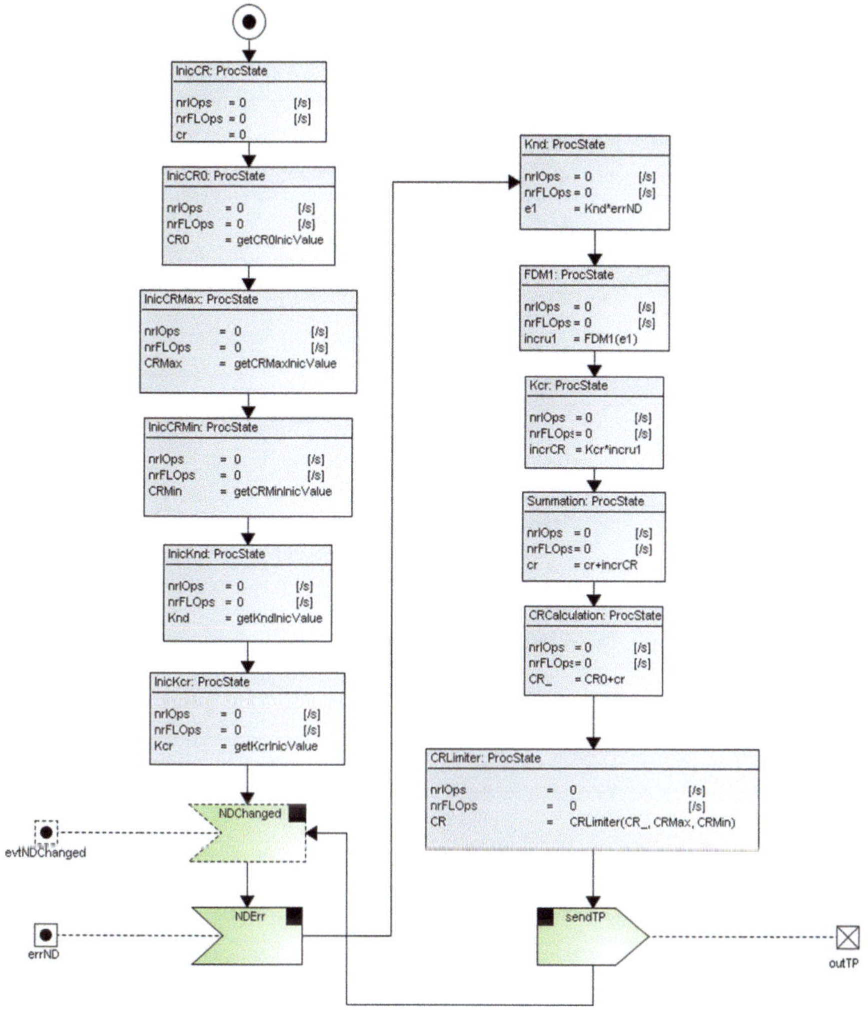

Fig. 6.13 Behavioral model for the *NDReasoning* task

The behavioral model for the *Secondary loop*, i.e. the one for the *BLReasoning* task, is provided in Fig. 6.14. Once the initial values for $\overline{ND}$, k_E and $k_{\Delta nd}$ have been set, the behaviour is quite similar to *NDReasoning* task. The task will wait for an incoming event denoting the node battery level has dropped below a critical level, defined at design time, and system reconfiguration might be needed. Once the event has been triggered, the task will wait for the difference between the critical and the current node battery level, provided by the *BLMonitor* task. The error will be amplified by k_E to achieve the desired reconfiguration dynamics. That will be the input to the FDM2 processing state, a fuzzy reasoner, which will actually make a decision and decide the increment to be applied to the current

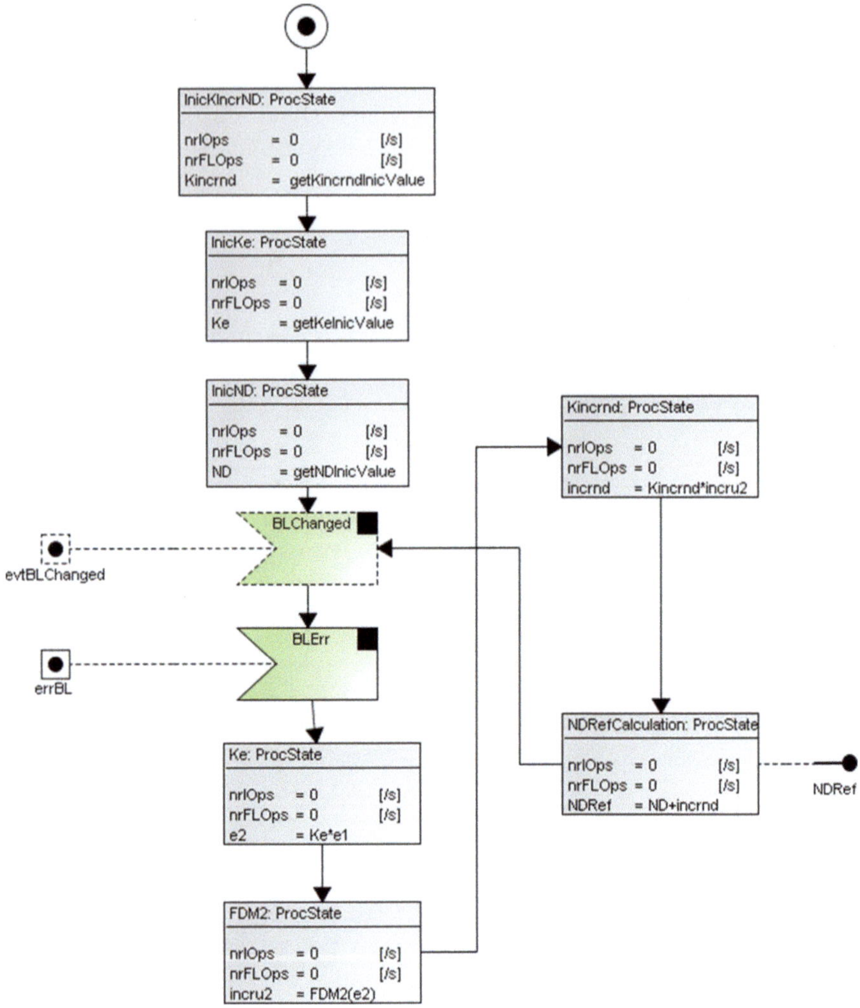

Fig. 6.14 Behavioral model for the *BLReasoning* task

node number of neighbors. The increment will be amplified by $k_{\Delta nd}$, which will also drive the reconfiguration dynamics. Finally, the result from the processing state *Kincrnd* will be added to $\overline{ND}$ to change the reference value for the desired node number of neighbors, an input reconfiguration parameter for the *ND_Monitor* task. Reducing the node number of neighbors will impact the network connectivity but will reduce the energy consumption as the power transmission would be also reduced.

6.3.3 *Physical Model*

6.3.3.1 Cargo Monitoring System: Physical Model

The physical model presents the real devices that will be part of the system as well as the communication channel among them. Both, devices and channels, could be described more in detail providing hardware and performance characteristics, such as the processor inside, the storage capabilities, the transmission speed and so forth.

The CMS physical model is depicted in Fig. 6.15. The sensor nodes deployed inside a container measure different physical magnitudes and they send the data to a gateway outside the container. That gateway relays the data to a back office computational infrastructure for further processing, using either a WiFi or 3G connection.

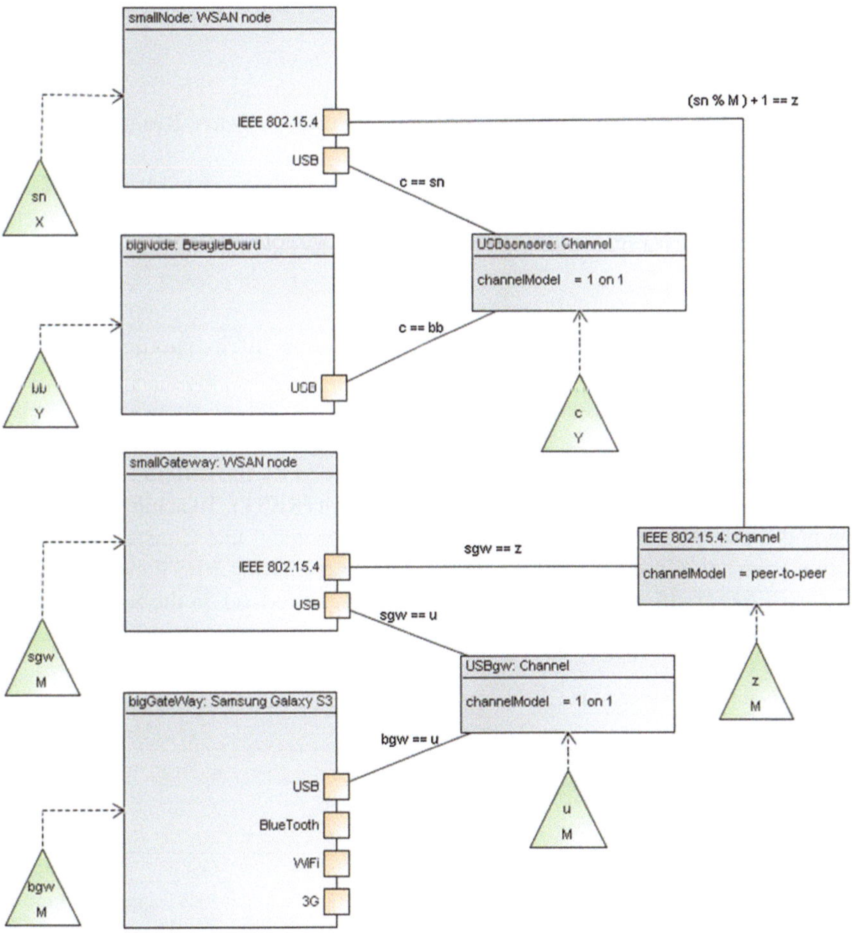

Fig. 6.15 CMS physical model

In Fig. 6.15, each sensor node is made up of two devices: an instance of the node class *BeagleBoard*, which models a BeagleBoard hardware platform, and an instance of the node class *WSAN node* (*small node* from now on), which models either a GNode, SunSPOT or TelosB hardware platform. The former has enough computational resources for running the core modules of the runtime services (middleware). The latter has sensors, wireless communication capabilities and basic computational resources for at least sensing and transmitting the measures. Each *BeagleBoard* is connected to a small node by means of a USB connection.

The gateways outside the containers are also made up of two devices: a Samsung Galaxy S3 smartphone (*bigGateway* from now on) and an instance of the node class WSAN node (*smallGateway* from now on). Both devices are connected by means of a USB connection. The smallGateway will interact with the sensor nodes over IEEE 802.15.4. The *bigGateway* will process the data received from the sensors through the *smallGateway* and will also keep a WiFi or 3G connection with the back office computational infrastructure.

6.3.3.2 Cargo Monitoring System: Hardware and Software Resources

The Table 6.3 summarizes the software resources running in each hardware platform.

The Figs. 6.16, 6.17 and 6.18 depict the hardware resources of the different elements in the physical diagram briefly described in the previous section.

6.3.3.3 Cargo Monitoring System: Communication Infrastructure

The wireless communications among the sensor nodes and the gateway outside the containers will be over Zigbee/IEEE802.15.4. The deliverable [3] suggests that the access to the capabilities and functionalities provided by the different modules of the middleware should comply with a SOA approach (REST). To achieve that, when the nodes and the gateway are communicating, a protocol like Constrained Application Protocol (CoAP) would be preferred for transporting REST messages over Zigbee/IEEE802.15.4, as it does not require a heavy overload on the sensor nodes.

Table 6.3 Software resources in CMS nodes

Hardware platform	Operating system	Execution platform
GNode	TinyOS	
TelosB	TinyOS	
SunSPOT	Squawk (JVM)	
BeableBoard	Ubuntu	OSGi
Samsung Galaxy S3	Android	REST application server

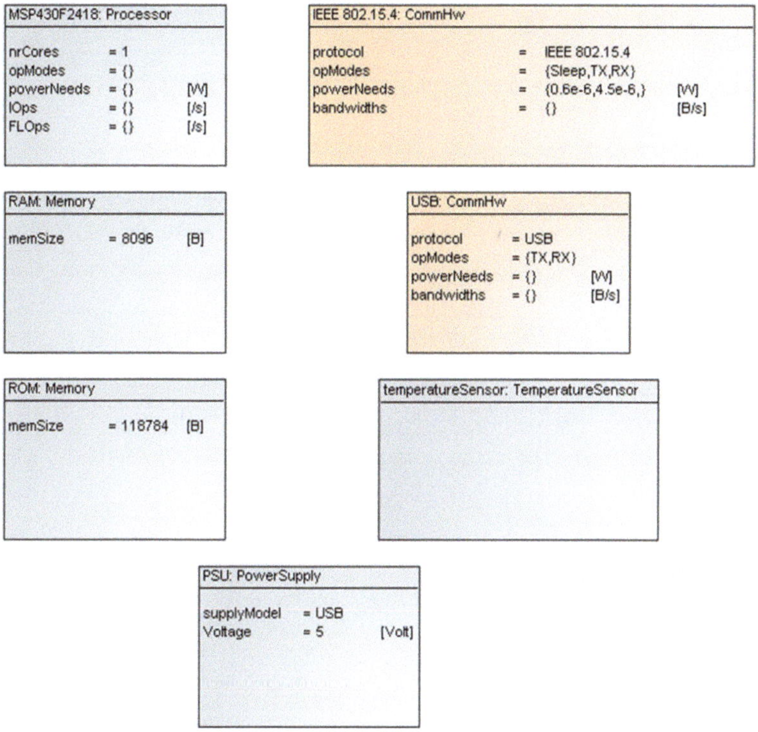

Fig. 6.16 Model for the GNode hardware platform

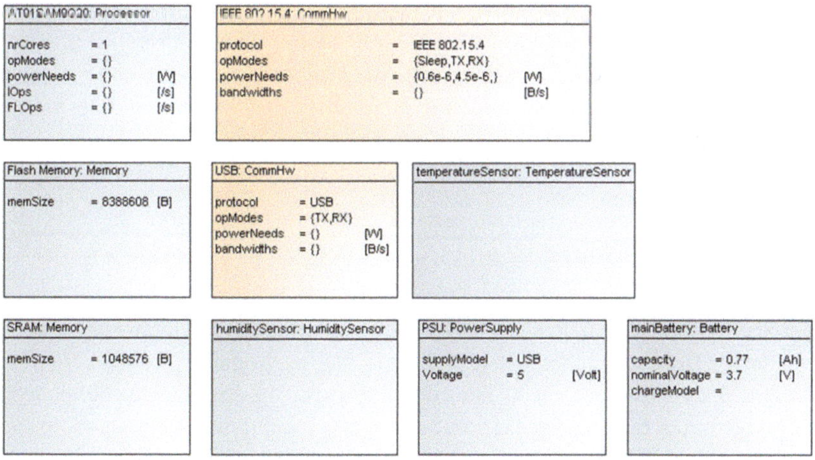

Fig. 6.17 Model for the SunSPOT hardware platform

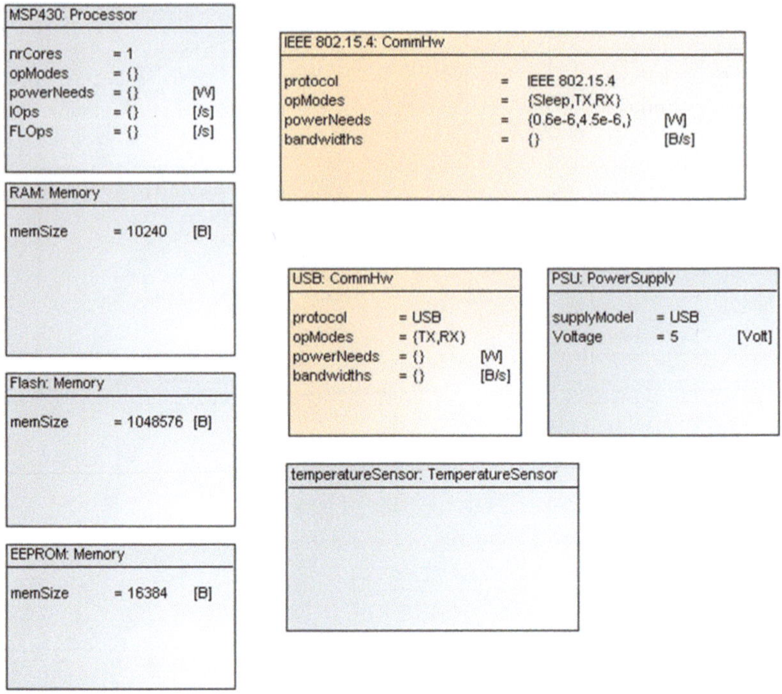

Fig. 6.18 Model for the TelosB hardware platform

The wireless communications among the gateways outside the containers and the back office computational infrastructure will be over WiFi or 3G connections. The computational resources of those communicating parts are enough for supporting the whole REST protocol stack when interacting with the middleware modules running in different devices.

6.3.4 Mapping Model

The mapping model establishes which tasks from the task model are to be run in which device from the physical model. It links the functionalities of the system and the device that will provide each. When defining a task, the designer can define which physical resources the task needs (processor, storage capabilities, etc.), and, to some extent, that might drive the mapping model as it would constrain which devices have enough resources to run specific tasks.

It has been already discussed that some tasks in the CMS will measure physical magnitudes and will achieve reconfiguration activities such as adapting the measures sampling rate, the state estimation algorithm, the way data is shared among nodes or the tasks that a node runs. Figure 6.19 depicts the mapping model of those tasks.

In the Fig. 6.19, the WSAN nodes carry two main activities: measure different physical magnitudes and estimate the container state. Such estimation could be based on shared data among the nodes, depending on the parameters ***share_sensor*** and ***share_state***, across the ports ***in_coop*** and ***out_coop*** over IEEE 802.15.4.

The BeagleBoard platforms store and retrieve the container state and run the reconfiguration task that will control the measured sampling rate, the way the container state is estimated and so on. There is another set of tasks that would be run in a BeagleBoard or in the gateway, depending on the parameter ***card_interpreter***: *stateInterpreter*, that analyses the container state and triggers events and alarms,

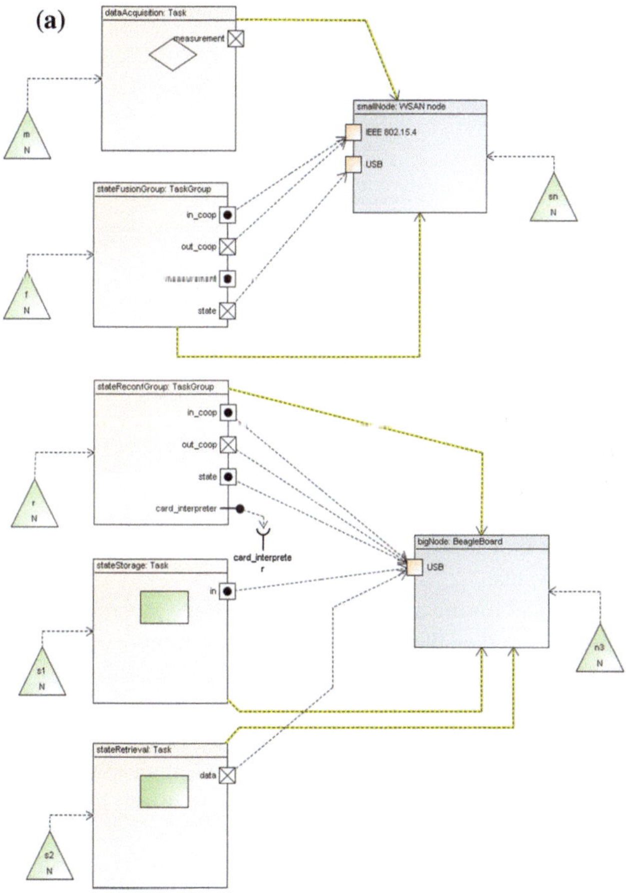

Fig. 6.19 **a** Mapping model for tasks reconfiguring container state estimation. **b** Mapping model for tasks reconfiguring container state estimation (cont.)

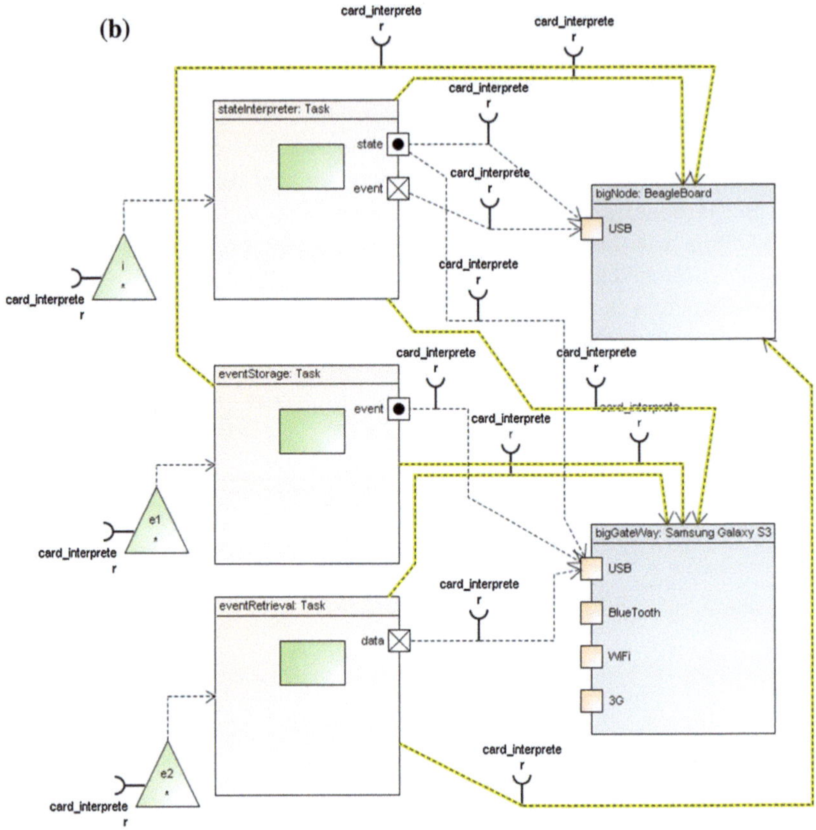

Fig. 6.19 (continued)

eventStorage, a task that store the events from *stateInterpreter*, and *eventRetrieval*, that retrieves past events.

The mapping model for the tasks that accomplish the reconfiguration of the node power transmission based on the battery level and the number of neighbors of a node is shown in Fig. 6.20.

The WSAN nodes, besides of measuring physical magnitudes, will adapt the node power transmission according to the expected battery lifetime and to the desired network connectivity. For performing all that, the WSAN nodes will host the tasks and tasks group *dataAcquisition*, *powerTransmissionControl* and *powerTramsissionReconfGroup*.

The gateway outside the container, a SamsungGalaxy S3 actually, will manage system alarms by making decisions on the events that are detected, as the Fig. 6.21 depicts. Those alarms will be pushed to the back office.

Other significant tasks groups that manage the self-adaptation capabilities concerning links quality, free storage and the battery lifetime, will be hosted in WSAN

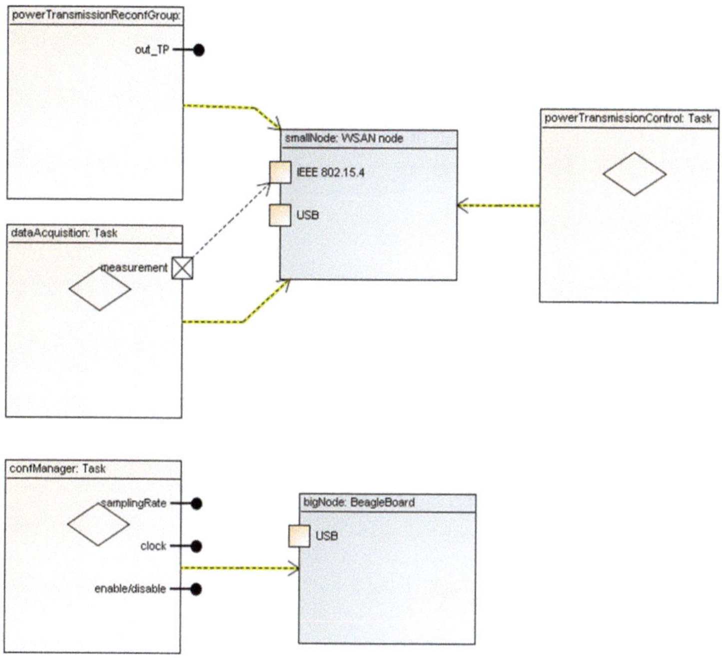

Fig. 6.20 Mapping model for power transmission reconfiguration

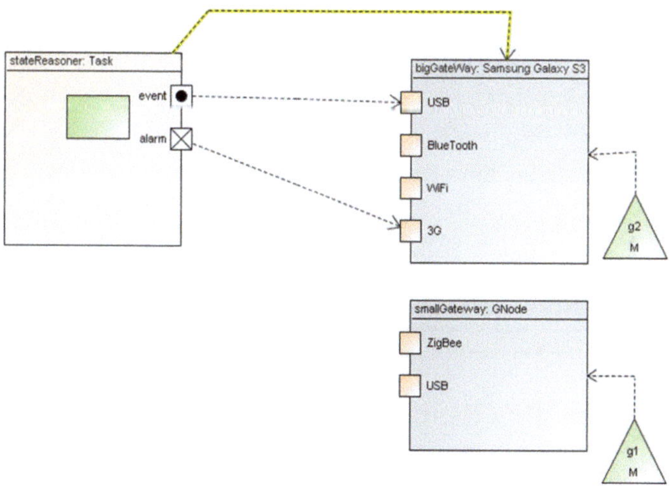

Fig. 6.21 Mapping model for the CMS gateway

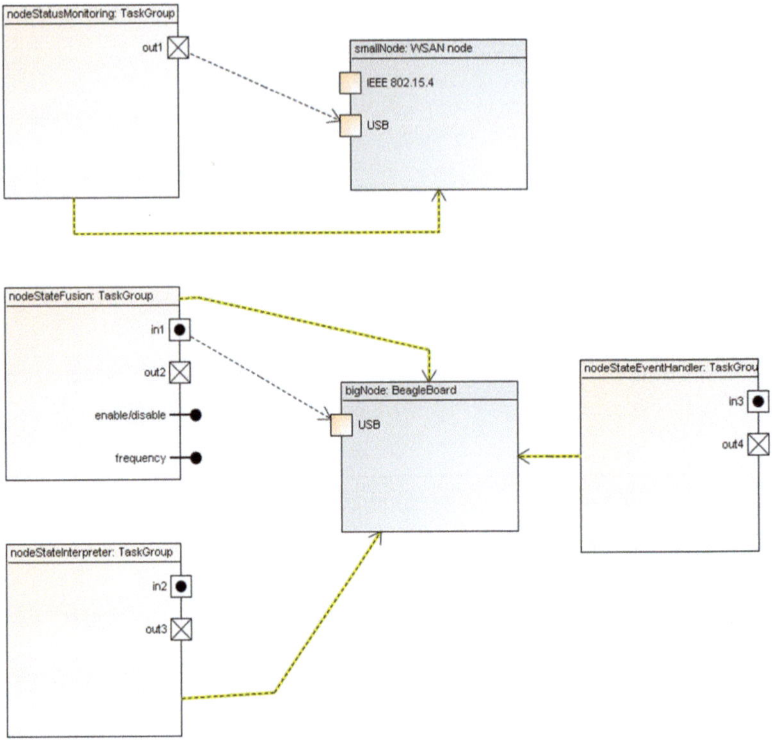

Fig. 6.22 Mapping model for tasks adapting links quality, free storage and battery level

nodes and BeagleBoard platforms as depicted in Fig. 6.22. The tasks group that monitors the communications links quality, the free storage and the battery level, will be hosted in a WSAN node. But the tasks groups that make decisions based on the monitored node state and manages events and alarms, will be hosted in BeagleBoard platform. The latter tasks groups are computing intensive and a WSAN node might not have enough computing resources.

6.4 Implementation Example

In this section an implementation example is presented according to the design described in the previous sections. In the implementation example a container is equipped with the cargo monitoring sensor system. In this example the sensor nodes measure the temperature and the humidity in the container and send the measurements and events to a server using a 3G uplink, which are visualized to the end user via a web interface.

In terms of adaptation the goal of the system is to optimize energy consumption while remaining operational according to the user requirements. This means the system can adapt the sleep/wake cycle to make sure the data reported to the user is updated more frequently when a potentially interesting event occurs, and less frequently when the battery levels starts to decrease.

In the meantime the communication to the server needs to stay operational even when 3G communication is not possible. Therefore a backup network is employed, simulating a situation of multiple containers sharing resources in a multi-hop network. When 3G connectivity drops for one gateway, the communication to the server is relayed via a network of other container nodes. This multi-hop network is supported by adaptive transmission power control, meaning that the intermediate nodes are communicating at power levels such that they reduce battery consumption and interference levels, while keeping the communication operational and reliable.

6.4.1 Implementation Hardware

In the CMS we used custom built sensor systems that consist of two nodes, a battery and a USB hub for attaching peripherals. The design layout is shown in Fig. 6.23 and the box itself is shown in Fig. 6.24. The main computing unit is a BeagleBone

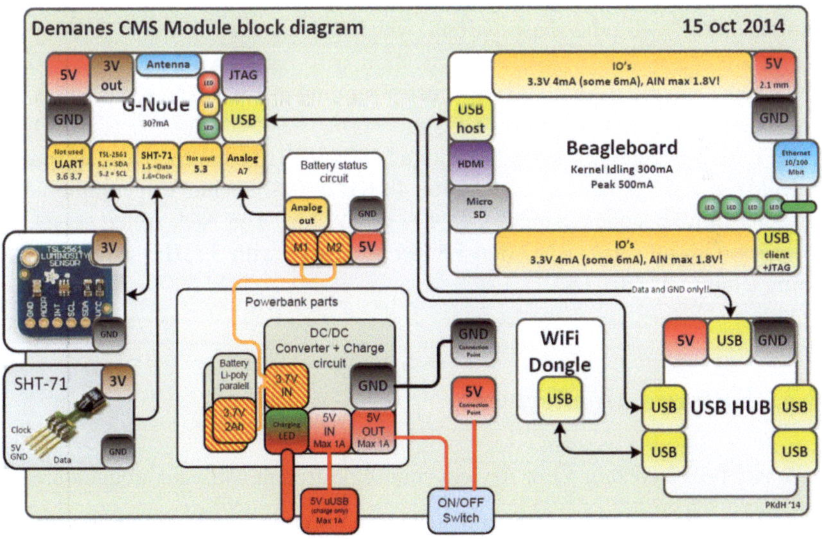

Fig. 6.23 The design specification of the CMS sensor box

Fig. 6.24 The CMS sensor box

Black[2] which has a 1 GHz AM3358BZCZ100 processor, 512 MB DDR3 800 MHz RAM and 4 GB of eMMC flash storage as well as a microSD card slot.

Attached to the beaglebone is a GNode[3] that is responsible for processing the sensor measurements. The GNode has a MSP430F2418 microcontroller processor, 116 KB of ROM, 8 KB of RAM and a CC1101 radio chip. The radio chip enables communication from one sensor system to another.

The gateways in the CMS also have a 3G dongle attached, as well as a SunSPOT basestation. The 3G dongle is used as the primary communication module with the back office server. In case the 3G is unavailable, the SunSPOTs are capable of communicating with other basestations, which are attached to other gateways, to form an alternative uplink for communication with the back office. SunSPOT are mobile nodes having a ARM920T processor running at 180 MHz, it has 512 KB of RAM and 4 MB of flash memory.

The TelosB devices are equipped with a MSP430F1611 microcontroller running at 4.15 MHz, and has 10 KB of RAM and 48 KB of flash memory. It communicates with other TelosB nodes using a CC2420 radio chip. The back office server is a webserver with an Intel Core2 duo E8400 processor with 3 GHz, 4GB of RAM, running Ubuntu 14.01 LTS.

6.4.2 Software Architecture

The sensor boxes contain a few devices running different software architectures.

- The BeagleBoards run a Debian distribution compiled by BeagleBoard and the uses the oracle java runtime environment 1.6.
- The GNodes run Contiki, and all code is implemented in C.

---

[2]http://elinux.org/Beagleboard:BeagleBoneBlack

[3]http://www.sownet.nl/download/G301Web.pdf

- The SunSPOTs run Java ME Squawk based on Java 1.4 directly on the processor.

The backoffice server is running Ubuntu 14.04.1 and uses MySQL 5.5.40 as the interface for data, and Apache 2.4.7 and PHP 5.5.9 for providing the web interface.

6.4.2.1 Middleware Modules

The Fig. 6.25 depicts the ANES middleware modules, also referenced in the Sect. 3.4 and fully described in [4]. The highlighted modules are those that have been used in the CMS.

The *Monitoring Network/Component Status* module has been used to log any CMS relevant message to monitor its activities and behaviour. The *Lifecycle* module has been used to manage CMS components lifecycle, i.e. to mainly start, stop and activate components. The *Parameterization* module supoorts the CMS reconfiguration parameters management. The *Reasoning Engine* provides a set of components that let developers deploy readily any kind of reasoner. The reasoner will get measures from any registered observer and can act on any registered actuator. A fuzzy-based reasoner and a PrologReasoner have been used in the CMS. Finally, the *Seamless Communication* module provides the means for forwarding messages through the SunSPOT network whenever the WiFi or 3G link is not availabel in a gateway.

The ANES middleware modules on the BeagleBoard have been implemented using the Apache Felix OSGi framework. We use the PrologReasoner to drive the adaptive reconfiguration, and a TimedTriggerPolicy to decide on system's configuration statically every 5 s.

Communication from one device to another is done using the CoAP. This is used for parameterizating the GNode from the BeagleBoard, gathering Observations and invoking Actions, and for logging the system state from the GNode in the Beaglebone.

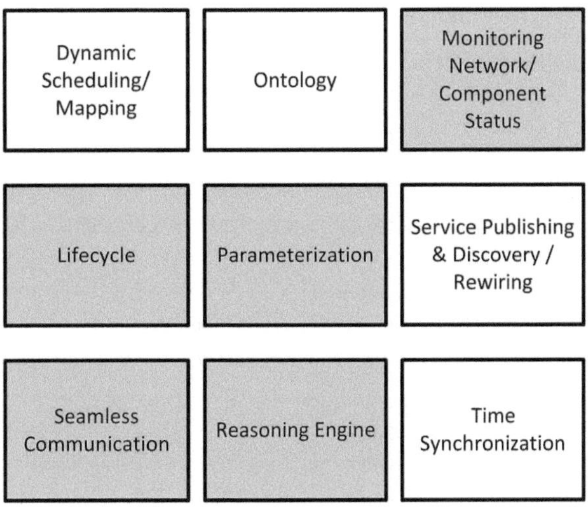

Fig. 6.25 ANES middleware modules

The seamless communication module is used for communicating from the Gateway to the server, either using a direct 3G uplink, or an alternative uplink via (a multi-hop network of) SunSPOTs.

In the application itself, parameterization and reconfiguration related functions were annotated using the middleware annotations, and the subsequent compatible applications were automatically generated during deployment

6.4.3 Use Cases

In the example implementation the behaviour of the sensor systems is shown as it monitors the container during shipment. In the demonstration the following scenarios have been demonstrated:

1. The battery of the sensor nodes run empty, and the system should reconfigure to save energy in the future.
2. The temperature or humidity gets outside of the expected conditions. Accordingly the sensor systems go into an increased awareness mode to increase the update rate.
3. The communication with the back office via 3G is disrupted, the system should reconfigure to use the backup SunSPOT uplink network.

To demonstrate these scenarios the sensor boxes are deployed in the container. In order to demonstrate the first scenario, the different sensor boxes emulate a different battery level. By doing so, different sensor boxes will configure to a different update round. E.g systems that have a relatively high battery level will communicate their state more often that systems that have a low battery.

For the second scenario an artificial increase in the temperature is induced. This results in a reported peak in the observer container temperature, and the corresponding node increases its update rate. By doing so it will increase the battery consumption rate, but also the rate at which information is reported to the gateway, the back-office and the end user.

For the final scenario the 3G dongle is removed from the gateway, at any point during the demonstration. The first next update to the back office will fail, and the gateway will try to send the cargo life update event (CLUE) via the SunSPOT network. If this process succeeds, the gateway at the other end of the SunSPOT network (which does still have an active 3G uplink) will send the CLUE to the back office.

6.4.4 Performance Considerations

In the cargo monitoring example the following KPI are used:

- Energy Consumption: The lifetime of the sensor nodes indicate the average energy consumption throughout deployment. Particularly the energy consumption of the sensor nodes and the SunSPOTs are of interest, because the gateway and the server lack the ability to reconfigure to decrease their energy consumption.

- Packet Delivery Ratio: The ratio of the delivered packets indicate the reliability of the communication network between different elements in the demo. Links of interest are the following: between the sensor nodes, between sensor nodes and the gateway, between the gateway and the back-office and finally the links between nodes in SunSPOT network.
- Packet Delivery Delay: The same links as for the Packet Delivery Ratio are of interest in terms of the delay they introduce.
- Memory Usage: The sensor boxes utilize the middleware, which induces a memory overhead. This overhead indicates the applicability of the middleware on devices with limited memory resources.
- Time to Adapt: The time needed for the system to adapt to a (new) situation shows the speed of adaptation. This time can be found through the logging of the nodes as it will show the time at which a system state is observed which requires reconfiguration, and the time at which the system is in a new configuration.

There are a few additional overheads that are induced by the middleware such as the overhead for parameterization, seamless uplink communication, or the overhead for CoAP protocol for interacting with the external devices.

The reconfiguration abilities of the system occurs in multiple levels. The reconfiguration of the sensor systems are driven by the PrologReasoner running on the BeagleBoard. The knowledge base of the reasoner defines two states the system can be in: a normal state, or an increased awareness state. When in the normal state the update rate of the sensors is either 30, 45 or 60 s depending on the battery level, and when in the increased awareness state it is 3, 5 or 10 s.

Using the log trace on the different nodes, we can see the reconfiguration taking place whenever the battery level changes to a different level, or when an (artificial) change in the temperature or humidity occurs. Also in these log files we can find the overhead of the middleware from these reconfiguration actions and observations.

Whenever a CLUE is sent to the back office, the system will try to send the message via the 3G uplink (since this has priority), and if it fails it will continue through alternative uplinks such as the SunSPOT network, until finally it will throw an exception. In the log files we can identify the events in which the 3G communication fails, and we can see that the middleware successfully uses the backup network to communicate to the server.

The adaptive power scaling method is driven by a Fuzzy logics control mechanism as schematically shown in Fig. 6.9. This mechanism makes sure that the minimal required transmission power is used to maintain a constant number of neighbors that can provide reliable communication. The required transmission power is reported to the gateway node via CoAP messages, and subsequently sent to the back office. Changes in the transmission power can thus be monitored via a web interface, when changes occur in the network topology.

6.5 Conclusions

The cargo state monitoring system described in this chapter has been designed and implemented using the design methodology, modeling language and the runtime services proposed in the previous chapters. The modeling language enabled the system architect to accommodate the necessary runtime reconfiguration requirements and simulate the system behaviour and other features in an early design phase. By using a simple hierarchy, the language enables gradual zoom-in into more and more detailed views, thus relieving the system architect from the need to create large models.

The runtime services and tools have proven a support when implementing the cargo monitoring system. The example implementation of a cargo state monitoring system has shown that the services enable quick prototyping, feature demonstration, simulation and decision making for the system architects.

From an application point of view the availability of runtime reconfiguration capabilities in the cargo state monitoring system ensures the system meets the demanding requirements of robustness, life-time and availability. The monitoring and control of battery depletion ensures the system lifetime throughout the container trip. Communication re-routing using an alternative network when the gateway connection is not available guarantees availability of the cargo state at all times. It also allows different cargo state monitoring systems to collaborate, e.g. when containers are stacked on a ship.

References

1. The DEMANES Consortium ARTEMIS JU-GA: 269334: D3.1, Modelling for architecture design: state of the art review and requirement analysis (2013). http://www.demanes.eu/sites/default/files/DEMANES_D3.1_PU_TNO_01.06_2013.07.02.pdf
2. The DEMANES Consortium ARTEMIS JU-GA: 269334: D3.6, Evaluation of fundamental algorithms for the demanes middleware (2013). http://www.demanes.eu/sites/default/files/DEMANES_D3.6_PU_TUD_03.00_20140107.pdf
3. The DEMANES Consortium ARTEMIS JU-GA: 269334: D4.3, Specification and software detail description of reference embedded software (2014). http://www.demanes.eu/sites/default/files/DEMANES_D4.3_02_00_20140114.pdf
4. The DEMANES Consortium ARTEMIS JU-GA: 269334: D4.5, Reference embedded sw: prototype and samples (2014). http://www.demanes.eu/sites/default/files/DEMANES_D4.5_PU_UPM_01_00_20141231.pdf

Index

Z. Papp and G. Exarchakos (eds.), *Runtime Reconfiguration in Networked Embedded Systems*, Internet of Things, DOI 10.1007/978-981-10-0715-6

MIX
Papier aus verantwortungsvollen Quellen
Paper from responsible sources
FSC® C105338

If you have any concerns about our products,
you can contact us on
ProductSafety@springernature.com

In case Publisher is established outside the EU,
the EU authorized representative is:
Springer Nature Customer Service Center GmbH
Europaplatz 3, 69115 Heidelberg, Germany

Printed by Libri Plureos GmbH
in Hamburg, Germany